Uncertainty

William Briggs

Uncertainty

The Soul of Modeling, Probability & Statistics

 Springer

William Briggs
New York City, NY, USA

ISBN 978-3-319-81958-7   ISBN 978-3-319-39756-6   (eBook)
DOI 10.1007/978-3-319-39756-6

Printed on acid-free paper

This Springer imprint is published by Springer Nature
The registered company is Springer International Publishing AG Switzerland

To my beautiful and patient wife.
It's about time.

Foreword

Quid est veritas? ("What is truth?") Even if detached from its religious roots, perhaps especially if detached from its religious roots, this is the most serious question a human being can ask. For if we do not have a comprehensible conception of what truth is, then we lack the foundation on which all statements must rest. While it is not the case that we presently speak only cacophony and write only nonsense, if our conceptions of certainty and uncertainty are murky, we then proceed with a disturbing absence of attachment of practice to justification.

And yet, for decades, indeed centuries, the accepted conception of truth has been one of skepticism, a denial that truth exists in more than a probabilistic manner. In this book, William Briggs challenges this accepted wisdom with powerful arguments explained most cogently. With airtight, deep logic, he exposes weaknesses of probability, statistics, causality, modeling, deciding, communicating, and uncertainty—the whole kit and caboodle, "everything to do with evidence."

This is no small claim, and I approached Briggs's work with some skepticism of my own. After all, our conceptions of probability, statistics, and the rest have seemed to work pretty well. Then I read Briggs's book.

Much of the first part of this book, indeed the gist before Briggs gets to work on his positive insights, is refutation of our accepted concepts of probability statistics, evidence, chance, randomness, regression analysis, parameters, hypothesis testing, and a host of other concepts insufficiently questioned until now.

All this sets the stage for Briggs's central point: All truths are known because of the conditions assumed. All probability, like all truth, is conditional. All truths that are known are known because of the conditions assumed.

Briggs brings to his work the widest range of relevant competencies. He has applied his extensive training and research to a wide range of analyses including cryptology, weather forecasting, prediction (and, perhaps most tellingly, the basic failure of prediction), and, more generally, philosophy of science and epistemology.

Briggs's central point is that truth exists, but in a world currently plagued by an over-certainty, which "is already at pandemic levels." It is the failure to understand that all probability is conditional on evidence and resides in the mind, not in objects—probability has no ontological existence—that makes this pandemic possible.

In practice, and in all of science, conditional truths are far more relevant than are necessary truths. While thought could not proceed without necessary truths (P is P and not P is not-P), it is probability and conditional truth that is the launch pad for Briggs's great many original thoughts and the arena that surrounds and binds them.

Uncertainty presupposes, and demonstrates, the existence of truth. Uncertainty must be about something. You cannot be uncertain about nothing. The something implied by uncertainty means that truth exists. Without truth there can be no probability. Since there is probability, there must be truth.

Despite this, probability, and more generally our conception of truth—indeed our conception of anything—must inevitably be anchored in a metaphysical ground. Our understanding of essence and our incomplete and often faulty knowledge of it make this inevitable

Probability is the central issue in this book. Beginning with the traditional definition of logic—the relationship between propositions, and with the separation of the logical from the empirical—Briggs emphasizes and exploits the fact that probability, too, concerns the relationship between propositions. "The rest," he writes, "is mere detail."

Probability is epistemologically conditional. It can be epistemologically true, but it does not exist in ontological reality, but in the epistemology of the mind. Unlike the moon, or the stars, or human beings, probability does not have an existence in reality.

Mathematical proofs depend on premises and chains of premises; proofs found to be incorrect are nearly always found so not on the basis of miscalculation, but of the failure to take into account a necessary constraint. (It is interesting that proofs shown to be incorrect are virtually always demonstrated to be incorrect for this reason and not because their conclusions are incorrect. The conclusions are virtually always shown to be correct by a later proof.)

This book is full of subtle surprises. For example, it is almost universally assumed that deductive proofs are certain, while inductive arguments are uncertain. "But because we know indubitable propositions more surely than any other, induction produces greater certainty than deduction."

As central as are Briggs's methodological insights, equally crucial are their implications for decision-making. Thus, for example, his suggestion that we eliminate hypothesis testing, which serves merely to affirm biases, would go far to improve the decisions resting on probability.

What I have written here is but a glance at the foundation on which Briggs's edifice rests. The deepest satisfaction to the reader of *Uncertainty* resides in following Briggs's thought and logic and the explanation they generate. Anyone who does so will find that this is a marvelous, marvelous book.

Professor Emeritus of Sociology   Steven Goldberg
City College, City University of New York
New York City, NY, USA
January 2016

Preface

Fellow users of probability, statistics, and computer "learning" algorithms, physics and social science modelers, big data wranglers, philosophers of science, epistemologists, and other respected citizens. We're doing it wrong.

Not completely wrong; not everywhere; not all the time; but far more often, far more pervasively, and in far more areas than you'd imagine.

What are we doing wrong? Probability, statistics, causality, modeling, deciding, communicating, uncertainty. Everything to do with evidence.

Your natural reaction will be—this is a prediction based on plentiful observations and simple premises—"Harumph." I can't and shouldn't put a numerical measure to my guess, though. That would lead to over-certainty, which I will prove to you is already at pandemic levels. Nor should I attempt to quantify your *harumphiness*, an act which would surely contribute to scientism, which is when pseudo-numerical values assigned to mental states are taken as scientific.

Now you may well say "Harumph," but consider that there are people who think statistical models prove causality or the truth of "hypotheses," that no probability can be known with certainty until the sound of the last trump, that probabilities can be read from mood rings, that induction is a "problem," that randomness is magic, that chance is real, that parameters exist, that *p*-values validate or invalidate theories, that computers learn, that models are realer than observations, and that model fit is more important than model performance.

And that is only a sampling of the oddities which beset our field. How did we go awry? Perhaps because our training as "data scientists" (the current buzzword) lacks a proper foundation, a firm philosophical grounding. Our books, especially our introductions, are loaded with a legion of implicit metaphysical presumptions, many of which are false or which contradict one another. The student from the start is plunged into formula and data and never looks back; he is encouraged not to ask too many questions but instead to calculate, calculate, calculate. As a result, he never quite knows where he is or where he's going, but he knows he's in a hurry.

The philosophical concepts which are necessarily present aren't discussed well or openly. This is only somewhat rectified once and if the student progresses to the highest levels, but by that time his interest has been turned either to mathematics or to solving problems using the tools with which he is familiar, tools which appear "good enough" because everybody else is using them. And when the data scientist (a horrid term) finally and inevitably weighs in on, say, "what models really are," he lacks the proper vocabulary. Points are missed. Falsity is embraced.

So here is a philosophical introduction to uncertainty and the practice of probability, statistics, and modeling of all kinds. The approach is Aristotelian. Truth exists; we can know it, but not always. Uncertainty is in our minds, not in objects, and only sometimes can we measure it, and there are good and bad ways of doing it.

There is not much sparkling new in this presentation except in the way the material is stitched together. The emphasis on necessary versus local or conditional truth and the wealth of insights that brings will be unfamiliar to most. A weakness is that because we have to touch on a large number of topics, many cannot be treated authoritatively or completely. But then the bulk of that work has been done in other places. And a little knowledge on these most important subjects is better than none, the usual condition. Our guiding light is Thomas Aquinas, *ora pro nobis*, who said, "The smallest knowledge that may be obtained of the highest things is more desirable than the most certain knowledge obtained of lesser." It is therefore enough that we form a fair impression of each topic and move onward. The exceptions are in understanding *exactly* what probability is and, as importantly, what it is *not* and in comprehending just what models are and how to tell the good from the bad.

This isn't a recipe book. Except for simple but common examples, this book does not contain the usual lists of algorithms. It's not that I didn't want them, it's more that many proper ones don't yet exist or aren't well understood; and anyway, they can be a distraction. This book is, however, a guide on how to create such recipes and lists, as well as a way to shoehorn (when possible) older methods into the present framework when new algorithms haven't yet been created. This book is thus ideal for students and researchers looking for problems upon which to work. The mathematical requirements are modest: This is not a math book. But then probability is not a mathematical subject, though parts of it are amenable to calculation.

Some will want to know what to call this unfamiliar new theory. Well, it isn't a theory. It is the way things are. The approach taken is surely not frequentist, a method which compounds error upon error, but it is also not Bayesian, not in the usual sense of that term, though it is often close in spirit to objective Bayesianism. There is no subjectivism here. The material here is closely aligned to Keynes's, Stove's, and Jaynes's logical probability. Many elements from the work of these and similar gentlemen are found here, but there are also subtle and important differences. If a name must be given, probability as argument is as good as any, though I prefer simply probability.

If we're doing it wrong, what's right? Return to understanding cause. Models should be used to make probabilistic predictions of observable entities. These predictions can, in turn, be used to make decisions. If the predictions fail, the models fail and should be abandoned. Eliminate all forms of hypothesis testing, Bayesian or frequentist, and forever banish p-values, which only serve to confirm biases. Do not speak of parameters; talk of reality, of observables. This alone will go miles toward eliminating the so-called replication crisis.

Here is the book in brief. All truth is conditional on or with respect to something. There are thus necessary or universal and conditional or local truths. Truth resides in the mind and not in objects except in the sense that objects exist or not. Truth is not relative in the modern sense of that word. Probability aims at truth. We come to know many truths via induction, which is widely misunderstood and is not a "problem"; indeed, it provides the surest form of knowledge. Logic is the study of the relationship *between* propositions and so is probability. All probability, like all truth, is therefore known because of the conditions assumed.

Most probabilities are not quantifiable, but some are. Probability is not subjective, and limiting relative frequency is of no use to man or beast. Chance and randomness are not mystical entities or causes; they are only other words for ignorance. Science is of the empirical. Models—whether quantum mechanical, medical, or sociological—are either causal or explanative. Causal models, which are in reality as rare as perfect games (baseball, of course), provide certainty, and explanative models state uncertainty. Probabilistic models are thus not causal (though they may have causal elements).

Bayes is not what you think. Hypothesis testing should immediately and forever be tossed onto the scrap heap of intellectual history and certainly never taught to the vulnerable. Probability is not decision. The parameter-centric, even parameter-obsessed, way of thinking about models must also be abandoned; its use has led to widespread, enormous over-certainty and caused more than one soul to be lost to scientism. Its replacement? If somebody asks, "How does changing X change my uncertainty in Y" *tell them that* and nothing else. Models, which provide the basis of these statements, are and must be checked against reality. The best way to check against reality is conditional on the decisions to which models are put. The most common, widespread errors that come in failing to not treating probability logically are shown, including the common mistakes made in regression, risk measures, the overreliance on questionnaires, and so on.

The language used in this book will not be familiar to regular users of probability and statistics. But that is rather the point. It ought to be. Along the way we'll solve things like induction, Gettier "problems," Grue, the Doomsday Argument, so-called paradoxes in probability assignment, the reproducibility crisis, and much more.

How working statisticians and probabilists should read this book. Start with Chap. 6 "Chance and Randomness," and then read the four successive chapters "Causality," "Probability Models," "Statistical and Physical Models," and "Modeling Strategy and Mistakes." After this, return to the beginning for the proofs of the assumptions made in those chapters.

Everybody else, and in particular students, should start at the beginning.

Manhattan Island, New York City, NY, USA   William Briggs
December 2015

Acknowledgments

Pretium non pro veritate.

The motivation behind the tone and structure of this book can be summarized in the old saying (slightly modified), "The love of theory is the root of all evil." The text itself is owed to four men whom I have never met: Aristotle, St. Thomas Aquinas, David Stove, and E.T. Jaynes. My task has been to take what these geniuses have provided and to synthesize out of it a philosophy of uncertainty for an age over-devoted to the appearance of certainty. As for originality, my book stands in the same class as Jeeves's suggested work *The Children's Book of American Birds*, which he had, through an unknown impecunious author in the name of a certain Corky, dedicated to one Alexander Worple, so that Bertie Wooster's pal Corky, who was the nephew of this Worple and who was on a tight allowance controlled by his uncle, could ingratiate himself to Worple, the selfsame Alexander Worple who was the author of the acclaimed *American Birds* and *More American Birds*.

I thank Russell Zaretzki, with whom I discussed and developed some of the material in Chap. 8 on the problem and origin of parameters. Thanks also to Christopher Monckton, 3rd Viscount Monckton of Brenchley, Willie Soon, and David Legates, all of whom have contributed to the material on time series in the final chapter. I thank Samuel Bacharach at Cornell University for providing me employment while writing. John Watkins read through the manuscript and provided helpful comments. Finally, I thank the dedicated readers of my blog, who have encouraged me and sharpened my thinking over the years. A book page at my site is available for updates, homework, and so forth: Go to wmbriggs.com/book.

Steven Goldberg is credited with the book's title. I had run through dozens of possibilities, all of them clunky and cumbersome or narrow and misleading.

Contents

Chapter 1
Truth, Argument, Realism

"Quid est veritas?"

The answer to the above, perhaps the most infamous of all questions, was so obvious that Pilate's interlocutor did not bother to state it. Truth was there, in the flesh, as it were, and utterly undeniable. Everyone knows the sequel. Since that occasion, at which the answer was painfully obvious, the question has been re-asked many times, with answers becoming increasingly skeptical, tortured, and incredulous. The reasons for this are many, not the least of which is that denial of truth leads to interesting, intellectually pleasing, unsolvable but publishable puzzles.

Skepticism about truth is seen as sophistication; works transgressive to truth are rewarded, so much so that finding an audience accepting of truth is increasingly difficult. More than sixty years ago Donald Williams [224], exasperated over the pretended academic puzzlement over the certainty of truth, said the academy

> in its dread of superstition and dogmatic reaction, has been oriented purposely toward skepticism: that a conclusion is admired in proportion as it is skeptical; that a jejune argument for skepticism will be admitted where a scrupulous defense of knowledge is derided or ignored; that an affirmative theory is a mere annoyance to be stamped down as quickly as possible to a normal level of denial and defeat.

Yet truth is our goal, the only destination worth seeking. So we must understand it. There are two kinds of truth: ontological and epistemological, comprising existence and our understanding of existence. Tremendous disservice has been done by ignoring this distinction. There are two modes of truth: necessary and local or conditional. From this seemingly trivial observation, everything flows.

© Springer International Publishing Switzerland 2016
W. Briggs, *Uncertainty*, DOI 10.1007/978-3-319-39756-6_1

1.1   Truth

Truth exists, and so does uncertainty. Uncertainty acknowledges the existence of an underlying truth: you cannot be uncertain of nothing: nothing is the complete absence of *any*thing. You are uncertain of something, and if there is some thing, there must be truth. At the very least, it is that this thing exists. Probability, which is the language of uncertainty, therefore aims at truth. Probability presupposes truth; it is a measure or characterization of truth. Probability is not necessarily the quantification of the uncertainty of truth, because not all uncertainty is quantifiable. Probability explains the limitations of our knowledge of truth, it never denies it. Probability is purely epistemological, a matter solely of individual understanding. Probability does not exist *in* things; it is not a substance. Without truth, there could be no probability.

Why a discussion of truth in a book devoted to probability? Since probability is the language of uncertainty, before we can learn what it means we need to understand what it is that probability aims at. Hempel understood this, but couldn't help himself from writing the word without scare quotes, as if "truth" might not exist, [110]. What is the nature of probability's target? What does it mean to be uncertain? How do we move from uncertainty to certainty? How certain is certain? It will turn out that statements of probability (assuming they are made without error, an assumption we make of all arguments unless otherwise specified) are true. When we say things like "Given such-and-such evidence, the probability of X is p", we mean to say either that (the proposition) X is true, or that not-X is. So truth must be our foundation. What follows is not a disquisition on the subject of truth, merely an introduction sufficient to launch us into probability. This chapter is also a necessity because the majority of Western readers have grown up in a culture saturated in relativism. There is ample reason Pilate's question is so well remembered.

Our eventual goal is to grasp models, and models of all kinds, probabilistic or otherwise, are ways of arguing, of getting at the truth. All arguments, probabilistic or not, have the same form: a list of premises, supposeds, accepteds, evidence, observations, data, facts, presumptions, and the like, and some conclusion or proposition which is thought related to the list. Related how and in what way is a discussion that comes later, but for now it loosely is associated with what *causes* the proposition to be true. Arguments can be well or badly structured, formally valid or invalid, and sound or unsound. Unlike most logical, mathematical, and moral arguments, which often end in truth, probabilistic arguments do not lead to certainty. Whenever a probabilistic argument is used, it is an attempt to convince someone how certain a proposition is in relation to a given body of evidence, and *only that* body of evidence.

Anybody who engages in any argument thus accepts that certainty and truth exist. We should have no patience for philosophical skepticism, which is always self-defeating. If you are certain there is no certainty, you are certain. If it is true that there is no truth, it is false there is no truth. If you are certain that "Every proposition is subject to uncertainty" then you speak with forked tongue. Certainty and truth therefore exist. But we must understand that truth resides in our intellects and not in objects themselves, except in the sense of existence. That being so, probability also does not exist physically; it also resides in our intellects and not in things themselves.

All arguments have stated and tacit premises, with those tacit usually about the meaning of the words and grammar used to state the argument, but also about how arguments themselves are to be interpreted, about how we move from premise to conclusion. Confusion usually enters when there are misunderstandings or disagreements formed about the tacit premises. Badly structured arguments are incautious in their use of tacit premises, containing too many or those which are prone to dispute. Ajdukiewicz confirms this in his lost classic *Pragmatic Logic*, an excellent book for students to understand the nature of arguments, [4]. There is also a burgeoning field called *argumentation theory* which can be looked up.

1.2 Realism

No definition of truth is better or more succinct than Aristotle's: "To say of what is that it is not, or of what is not that it is, is false, while to say of what is that it is, and of what is not that it is not, is true." St Thomas Aquinas, following Aristotle's *Metaphysics*, in his *Summa Theologica* (First Part, Q. 16, articles 1 and 8) said "the true denotes that towards which the intellect tends." "Truth, properly speaking, resides only in the intellect, as said before (1); but things are called true in virtue of the truth residing in an intellect."

This view encapsulate what is called *correspondence* and reflect the metaphysics of (moderate) realism; see [222] and below in the chapter on Causality. When we later say of a proposition "It is necessarily true", this is never meant to imply that the proposition is true *in* or *because of* some theory. The proposition is necessarily true for reasons in the proposition itself and the evidence which supports it; the proposition is not true "in" or because of a theory. It is true because it is true.

Moderate realism is the common-sense position that there exist real things, that there is an existence independent of our minds, that an external world is "out there" and that we can know it, that we can "know things as they are in themselves", to coin a phrase. Moderate realism holds that greenness exists apart from or in addition to individual green things; exists as an intellectual idea, that is. Realism says the idea of color exists independent of individual colored things. Mathematicians are realists when they insist all triangles have three straight sides and an interior sum of angles of 180°. Individual approximations to or implementations of triangles also exist, but given the way the world is, all are imperfect representations of the universal ideal. Try drawing one. Catness exists and so do individual cats. We can tell cats from dogs because we know the nature or essence of both. Knifeness exists as do individual knives, even though it's not always clear if a given object is a knife or only acts like one.

These natures (or ideas) are universals. They don't exist as physical objects in some ethereal realm, *à la* Plato; instead they exist in the objects which instantiate them—redness exists in red apples, knifeness exists in cleavers—or they exist as idea in intellects, as immaterial concepts. This is scholastic realism, a modified form of Aristotle's philosophy. Excellent introductions to moderate realism are given by Feser [70, 71].

Contrary to realism is nominalism, which denies universals exist. Under this view, individual triangles exist but there is no concept of an ideal, perfect triangle. This appears to leave out mathematical definitions and, it would seem to follow, all of mathematics, since this field is founded on universal truths (see below; also see Franklin's Aristotelian conception of mathematics, [81]). Under nominalism, two drawings of triangles are not two drawings of triangles, just two drawings which might have vague similarities, the similarities bespeaking of no central thing in common. How, then, if nominalism is true, could we even have the word *triangle* or even *similarity*? *Man* is also therefore a meaningless term: there are individual bipedal creatures which might coincidentally look somewhat alike and share some DNA (but is all DNA actually DNA?), just as they are more dissimilar to quadrupedal creatures. The higher concept of man or human being holds no higher meaning. Things do not instantiate natures. Things just *are,* never mind how. Most working scientists are not nominalists, for obvious reasons.

Nominalism comes in various forms and subtleties, but no branch holds any interest for probability and statistics. If there were no universals, there would be little point in conducting experiments or grouping data, which admits of universals or essences. The acts of grouping and collating say, do they not?, "All these data represent the same underlying essential thing." Even those dismal objects *p*-values admit of universal "null" and "alternate" hypotheses; these surely bespeak of universal essences and do not point to physical substances (*p*-values, God rot them, are discussed later). And neither is probability, as de Finetti taught us in a loud voice, a tangible physical quantity, something that can be measured with a physical apparatus. Probability, like logic, as we'll see, assumes universals.

The opposite of nominalism (if such a thing could have an opposite) is idealism, the concept that reality does not exist, rather that individual physical objects do not exist, but that only universals do. Our thoughts are capital-I It, our thoughts are everything, our thoughts *define* existence. If so, how do we know when you and I are thinking of the same thing? We cannot. I don't consider idealism to be on any interest. The best overview and refutation of idealism is found in David Stove's essay "Idealism: A Victorian horror story", [208].

There are many other ways for thought to go wrong, and those which have a bearing on probability will be outlined later. For now, I'll boldly state all scientists are realists, or ought to be. There's no use for a scientist who subscribes to some form of idealism. After all, if the universe is only in his mind, there's no guarantee that the universe which is my mind is in any way the same thing as the universe in his. If idealism is true, why not make up how the universe is? Saves research time. If nominalism is true, what is true here might not be true there, and it is of little to no use to speak of "laws" or causes.

1.3   Epistemology

Can we know any truths? Yes. And if you disagree you necessarily agree. In disagreeing you'd at least know that you can't know anything, which would be a truth, and then you'd realize you bit yourself in the tail. Any attempt to deny there are truths is self-contradictory. Roger Scruton said that the people who tout theories which deny truth or our knowledge of it are inviting us to disbelieve them, an invitation which we eagerly accept, [193].

That there are truths and we can know them is traditionally called rationalism. A prime example of a known truth is Aristotle's principal of non-contradiction. The epistemic version states that a proposition cannot be both true and false simultaneously (given the same evidence). It is impossible, and not just unlikely, for somebody to doubt this principle. It is possible, and unfortunately not uncommon, for some to *claim* to doubt it. But claiming and doing are not identical as everybody knows, and that is why we have the words like *deception, mistaken,* and *lying*—words, incidentally, which admit the existence of truth and knowledge. Claiming to doubt the principle of non-contradiction is like the man who boasts of disbelieving the reality of gravity. No matter the degree of his earnestness or the number of his scholarly credentials, if he takes a long walk off a short dock he is going to end up wet.

A ontological version of the non-contradiction principle is that something cannot be and not-be at the same time, that something cannot exist and not-exist simultaneously. Existence is an ontological truth. You cannot exist and not-exist at the same time; further, it is impossible, and not just unlikely, to believe that you exist and that at the same time don't exist. This is not the same as saying, for example with respect to certain very small objects in physics, that you do not *know* if or where a thing exists or not. A thing's existence and our knowledge of it are different. Indeed, the mixing up of epistemological and ontological claims is a routine problem in probability.

Everyone, regardless of what they might claim, knows that an external world exists. And all scientists ought to admit it, else they're in the wrong business. This is another way to state realism. Anybody asking the question of another, "Does an external world exist?" has answered it affirmatively, since to ask it requires a person to ask and another to answer it, hence an external world in which the other person exists to answer it, hence we can know it exists, hence we know there are other people, too (the traditional way to phrase it is that we know there are "other minds").

Another truth known to everybody is that solipsism is impossible. Again, if you disagree with me, you agree with me and acknowledge the complete fallaciousness of your position because, of course, to disagree with *me* implies someone other than yourself exists, hence solipsism is false.

But what if I were an illusion? What if, that is, you were hallucinating my obstreperousness? From David Stove's masterful essay "I only am alone escaped to tell thee: Epistemology and the Ishmael Effect", [207, pp. 61–82]:

[I]t is true, and also contingent, that some of us sometimes hallucinate. But it does not follow from that, (even if Descartes thought it did), that it is logically possible that all of us are always hallucinating. Some children in a school-class may happen to be below the average level of ability of children in that class, but it not logically possible that all of them are. Neither is it logically possible that *we* are all always hallucinating. For we—that is, all human beings—are perceived by (unless indeed we are hallucinations of) at least one human being: ourselves if no other. Whence, on the supposition that we—that is, all human beings—are *always* hallucinating, it follows that all human beings are hallucinations of at least one human being. And that is not logically possible.

Empiricism insists on the observational verifiability of all propositions, in contrast with realism, which does not. But not all propositions can be verified; think, for example, of truths reached by induction or mathematical deduction, especially statements about various infinite sets and so forth. The realism-rationalism view says all knowledge begins in sense impressions, and then moves from those particulars to grasp universals, which are entities which cannot be checked or verified empirically. Since most of our reasoning involves these universals, we'd be in a world of intellectual hurt if empiricism were true. *Only* the experiments we have seen are those we can say are true. We could never extrapolate from them. Those results which are merely similar or the same at other times and places might, under strict empiricism, be different. Mathematical axioms cannot be seen, touched, tasted, heard, or smelt, yet we insist on their truth. That logic in particular cannot be wholly empirical is dealt with in the next chapter, a useful exercise because probability follows directly from logic.

1.4 Necessary and Conditional Truth

Given "x, y, z are natural numbers and $x > y$ and $y > z$" the proposition "$x > z$" is true (I am assuming logical knowledge here, which I don't discuss until Chap. 2). But it would be false in general to claim, "It is true that '$x > z$'." After all, it might be that "$x = 17$ and $z = 32$"; if so, "$x > z$" is false. Or it might be that "$x = 17$ and $z = 17$", then again "$x > z$" is false. Or maybe "$x =$ a boatload and $z =$ a humongous amount", then "$x > z$" is undefined or unknown unless there is tacit and complete knowledge of precisely how much is a boatload and how much is a humongous amount (which is doubtful). We cannot dismiss this last example, because a great portion of human discussions of uncertainty are pitched in this way.

Included in the premise "x, y, z are natural numbers and $x > y$ and $y > z$" are not just the raw information of the proposition about numbers, but the tacit knowledge we have of the symbol >, of what "natural numbers" are, and even what "and" and "are" mean. This is so for any argument which we wish to make. Language, in whatever form, must be used. There must therefore be an understanding of and about definitions, language and grammar, in any argument if any progress is to be made. These understandings may be more or less obvious depending on the argument. It is well to point out that many fallacies (and the best jokes) are founded on equivocation, which is the intentional or not misunderstanding double- or multiple-meanings of words or phrases. This must be kept in mind because we often talk about how the mathematical symbols of our formulae translate to real objects, how they matter to real-life decisions. A caution not heard frequently enough: just because a statement is mathematically true does not mean that the statement has any bearing on reality. Later we talk about how the deadly sin of reification occurs when this warning is ignored.

We have an idea what it means to say of a proposition that it is true or false. This needs to be firmed up considerably. Take the proposition "a proposition cannot be both true and false simultaneously". This proposition, as I said above, is true. That means, to our state of mind, there exists evidence which allows us to conclude this proposition is true. This evidence is in the form of thought, which is to say, other propositions, all of which include our understanding of the words and English grammar, and of phrases like "we cannot believe its contrary." There are also present tacit (not formal) rules of logic about how we must treat and manipulate propositions. Each of these conditioning propositions or premises can in turn be true or false (i.e. known to be true or false) conditional on still other propositions, or on inductions drawn upon sense impressions and intellections. That is, we eventually must reach a point at which a proposition in front of us just *is* true. There is no other evidence for this kind of truth other than intellection. Observations and sense impressions will give partial support to most propositions, but they are never enough by themselves except for the direct impressions. I explore this later in the chapter on Induction.

In mathematics, logic, and philosophy popular kinds of propositions which are known to be true because induction tells us so are called axioms. Axioms are indubitable—when considered. Arguments for an axiom's truth are made like this: given these specific instances, thus this general principle or axiom. I do not claim, and it is not true, that everybody knows every axiom. The arguments for axioms must first be considered before they are believed. A good example is the principal of non-contradiction, a proposition which we cannot know is false (though, given we are human, we can always *claim* it is false). As said, for every argument we need an understanding of its words and grammar, and, for non-contradiction specifically, maybe the plain observation of a necessarily finite number of instance of propositions that are only true or only false, observations which are consonant with the axiom, but which are none of them the full proof of the proposition: there comes a point at which we just believe and, indeed, cannot do other than *know* the truth. Another example is one of Peano's axioms. For every natural number, if $x = y$ then $y = x$. We check this through specific examples, and then move via induction to the knowledge that it is true for every number, even those we have not and, given our finiteness, cannot consider. Axioms are known to be true based on the evidence and faith that our intellects are correctly guiding us.

This leads to the concept of the truly true, really true, just-plain true, universally, absolutely, or the *necessarily* true. These are propositions, like those in mathematics, that are known to be true given a valid and sound chain of argument which leads back to indubitable axioms. It is not possible to doubt axioms or necessary truths, unless there be a misunderstanding of the words or terms or chain of proof or argument involved (and this is, of course, possible, as any teacher will affirm). Necessary truths are true even if you don't want them to be, even if they provoke discomfort, which (again of course) they sometimes do. Peter Kreeft said: "As Aristotle showed, [all] 'backward doubt' terminates in two places: psychologically indubitable immediate sense experience and logically indubitable first principles such as 'X is not non-X' in theoretical thinking and 'Good is to be done and evil to be avoided' in practical thinking," [134, Part VI].

A man in the street might look at the scratchings of a mathematical truth and doubt the theorem, but this is only because he doesn't comprehend what all those strange symbols mean. He may even say that he "knows" the theorem is false—think of the brave soul who claims to have squared the circle. It must be stressed that this man's error arises from his not comprehending the whole of the argument. Which of the premises of the theorem he is rejecting, and this includes tacit premises of logic and other mathematical results, is not known to us (unless the man makes this clear). The point is that if it were made plain to him what every step in the argument was, he *must* consent. If he does not, he has not comprehended at least one thing or he has rejected at least one premise, or perhaps substituted his own unaware. This is no small point, and the failure to appreciate it has given rise to the mistaken subjective theory of probability. Understanding the *whole* of an argument is a requirement to our admitting a necessary truth (our understanding is obviously not required of the necessary truth itself!).

From this it follows that when a mathematician or physicist says something akin to, "We now know Flippenberger's theorem is true", his "we" does not, it most certainly does not, encompass all of humanity; it applies only to those who can and *have* followed the line of reason which appears in the proof. That another mathematician or physicist (or man in the street) who hears this statement, but whose specialty is not Flippenbergerology, conditional on trusting the first mathematician's word, also believes Flippenberger's theorem is true, is not making (to himself) the same argument as the theory's proponent. He instead makes a *conditional* truth statement: to him, Flippenberger's theorem is *conditionally* true, given the premise of accepting the word of the first mathematician or physicist. Of course, necessary truths are *also* conditional as I have just described, so the phrase "conditional truth" is imperfect, but I have not been able to discover one better to my satisfaction. *Local* or *relative* truth have their merits, but their use could encourage relativists to believe they have a point, which they do not.

Besides mathematical propositions, there are plenty other of necessary truths that we know. "I exist" is popular, and only claimed to be doubted by the insane or (paradoxically) by attention seekers. "God exists" is another: those who doubt it are like circle-squarers who have misunderstood or have not (yet) comprehended the arguments which lead to this proposition. "There are true propositions" always delights and which also has its doubters who claim it is true that it is false. In Chap. 2 we meet more.

There are an infinite number and an enormous variety of conditional truths that we do and can know. I don't mean to say that there are not an infinite number of necessary truths, because I have no idea, though I believe it; I mean only that conditional truths form a vaster class of truths in everyday and scientific discourse. We met one conditional truth above in "$x > z$". Another is, given "All Martians wear hats and George is a Martian" then it is conditionally true that "George wears a hat." The difference in how we express this "truth is conditional" is plain enough in cases like hat-wearing Martians. Nobody would say, in a general setting, "It's true that Martians wear hats." Or if he did, nobody would believe him. This disbelief would be deduced conditional on the observationally true proposition, "There are no Martians".

We sometimes hear people claim conditional truths are necessary truths, especially in moral or political contexts. A man might say, "College professors are intolerant of dissent" and believe he is stating a necessary truth. Yet this cannot be a necessary truth, because no sound valid chain of argument anchored to axioms can support it. But it may be an extrapolation from "All the many college professors I have observed have been intolerant of dissent", in which case the proposition is still not a necessary truth, because (as we'll see) observational statements like this are fallible. Hint: The man's audience, if it be typical, might not believe the "All" in the argument means *all*, but only "many". But that substitution does not make the proposition "Many college professors are intolerant of dissent" necessarily true, either.

Another interesting possibility is in the proposition "Some college professors are intolerant of dissent," where *some* is defined as *at least one and potentially all*.[1] Now if a man hears that and recalls, "I have met X, who is a college professor, and she was intolerant of dissent", then conditional on that evidence the proposition of interest is conditionally true. Why isn't it necessarily true? Understand first that the proposition is true for you, too, dear reader, if we take as evidence "I have met X, etc." Just as "George wears a hat" was conditionally true on the other explicit evidence. It may be that you yourself have not met X, nor any other intolerant-of-dissent professor, but that means nothing for the epistemological status of these two propositions. But it now becomes obvious why the proposition of interest is not necessarily true: because the supporting evidence "I have met X, etc." cannot be held up as necessarily true itself: there is no chain of sound argument leading to indubitable axioms which guarantees it is a logically necessity that college professors must be intolerant of dissent. (Even if it sometimes seems that way.)

We only have to be careful because when people speak or write of truths they are usually not careful to tell us whether they have in mind a necessary or only a conditional truth. Much grief is caused because of this.

One point which may not be obvious. A necessary truth is just true. It is not true *because* we have a proof of it's truth. Any necessary truth is true *because* of something, but it makes no sense to ask why this is so for any necessary truth. Why

---

[1] I keep this definition throughout the book unless otherwise specified.

is the principle of non-contradiction true? What is it that *makes* it true? Answer: we do not know. It is just is true. How do we *know* it is true? Via a proof, by strings of deductions from accepted premises and using induction, the same way we know if any proposition is true. We must ever keep separate the epistemological from the ontological. There is a constant danger of mistaking the two. Logic and probability are epistemological, and only sometimes speak or aim at the ontological. Probability is always a state of the mind and not a state of the universe.

1.5 Science and Scientism

The example of the intolerant college professor is like most propositions in science. Examples, "Radium has the atomic weight of w", "The speed of light is c", "The earth is warmed by the sun's rays", "Creatures evolve by natural selection", and on and on. These statements are all *contingent*, meaning there is (so far) no known route to proving their necessary truth (though that is the goal in physics). They are all conditionally true, given various facts and evidence. In any of these propositions none of the conditioned-on facts or evidence meets the test of a sound chain of valid argument leading to indubitable axioms. In other words, none of these propositions are logically necessary. It is a logical possibility that any of them might be (necessarily or observationally) false. That radium does not have the atomic weight of w might be false if the equipment, no matter how sophisticated or fine, erred in its measurements. That the speed of light is some number might also be false for the same or some other reason. Many physical formulas (and this is obviously theory dependent) rely on "constants", such as the speed of light in a vacuum or Planck's constant, which are productions of the result of measurements. They are not themselves deduced from earlier truths; i.e. there is nothing which we know of that states Planck's constant must of logically necessity take the precise value it does (though, as I said, this is the goal of physics). The same is true for all statistical use of parameters. This lack spoken of is what makes these creatures *parameters*, about which much more later. That means any theory which relies on contingent premises might be false. It might be incredibly improbable, given the evidence we have, for our best scientific theories to turn out false, but we cannot claim *any* are necessarily true.

Scientific statements are therefore contingent statements which can only ever be conditionally true and not necessarily true (the math used *in* science is an exception, of course). All scientific propositions are therefore subject to doubt. Not always reasonable doubt. Here is a scientific proposition, "If I walk off the edge of the twenty story building I will fall." There is no chain of argument which proves this is universally true, therefore the proposition is contingent. It is not logically necessary that falling must occur. But I will not be walking off the edge of any twenty-story buildings. I'm also happy with the atomic weight of Radium, even though I've taken none of the pertinent measurements myself. The premise of trust is ever present, though as the business of science expands, this premise is weakened.

All science is an attempt to remove as much of the contingency as possible from the supporting evidence for propositions (theories) of interest. The ultimate Theory of Everything would be one which is necessarily true, which begins at indisputable axioms and progressed toward a complete explanation for how everything works, including complete deduced explanations of why the speed of light and Planck's constant (if they should turn out to remain finally important) take the values they do. Those who have read in physics know how distant, and perhaps even unattainable, this goal is.

People before Newton knew apples fell, and would say so. The reasons they gave for this produced conditional truths—"Apples fall because they have an affinity for the ground", maybe—which allowed for good predictions: sure enough, apples always fell. Nobody not delusional walked off a mountain cliff in *anno Domini* 1600 in expectation of not falling because they didn't understand Newton's theory of gravitation. Newton's great trick was to replace highly contingent and more-or-less dubious premises with better evidence which had less contingency. He never removed the contingency completely, of course. But then neither did Einstein when he refined Newton's premises further. And still nobody has supplied a universally true argument which shows the logical necessity of gravity behaving the way it does. Scientists labor still to remove the remaining contingencies (and there are plenty). Whether they can eliminate them *entirely* and arrive at scientific statements with all the rigor of mathematical proofs is not known. There is plenty of reason to doubt it, however; but that discussion would take us too far afield. Suffice to say that no known scientific theory is necessarily true. All are at best conditionally true, many are only probably true, and still others are probably or certainly false (examples of these will follow).

Many scientists, perhaps heeding too closely to their citizen cheering section, have the bad habit of insisting that their conditional truths are necessary truths. Some have the even worse habit of insisting probable truths are not only not conditionally but are universally true. Bad habits lead to iniquity, which in this case is the sin of *scientism*. This is the false belief that the only truths we have are scientific truths. Since scientific truths are only conditional at best, and likely only probable and sometimes false in fact—a truth captured in the slogan "science is self-correcting", which implies it errs—it is not possible that it is a necessary truth that conditional or probable truths are necessary truths. Tongue twisting? It is not from science we learn "I exist". Though, if it can be credited, some scientists would say that consciousness of our existence is an "illusion", an obviously self-contradictory proposition. Who is having the illusion? But that's not a problem for us to solve in a book on probability. Science is also mute on all mathematical (necessary) truths, which is amusing because scidolators (those who inveterately practice scientism) often wield mathematical truths to show how scientific they are.

Jacques Barzun said this about scientism, [14]: "Scientism is the fallacy of believing that the method of science must be used on all forms of experience and, given time, will settle every issue." And Pascal in his *Pensees* had this to say, an observation which could be the motto of this book:

> The world is a good judge of things, for it is in natural ignorance, which is man's true state. The sciences have two extremes which meet. The first is the pure natural ignorance in which all men find themselves at birth. The other extreme is that reached by great intellects, who, having run through all that men can know, find they know nothing, and come back again to that same ignorance from which they set out; but this is a learned ignorance which is conscious of itself. Those between the two, who have departed from natural ignorance and not been able to reach the other, have some smattering of this vain knowledge, and pretend to be wise. These trouble the world, and are bad judges of everything. The people and the wise constitute the world; these despise it, and are despised. They judge badly of everything, and the world judges rightly of them.

The increasing politicization of science is also distressing. This is found whenever is heard somebody (almost never himself a scientist) screeching (this is never spoken politely) about some contingent proposition, "The debate is over!", as if their frenzy or level of ardency removed the obvious contingencies from the proposition in dispute. This tactic is always an obvious fallacy (unless it is applied to a demonstrable necessary truth). But this subject is too depressing to continue, so let it pass. We later meet many examples of scientism.

1.6   Faith

Faith is another difficult word. It has connotations of trust and honesty, but also of religion. In religion it is used to describe a kind of belief or as a label for a system or practice, e.g. "the Methodist faith." But you'll have noticed I used it above when describing epistemology. It is not out of place. To repeat: we know axioms and the like are true because our intellects tell us they are, and we *trust* that our intellects are not misleading us; that is, we have *faith* in our inductions. Faith is in this sense ultimate belief, the ground of all our beliefs. Belief is a *decision*, an act on top of knowledge or uncertainty. We prove via induction an axiom is true. This is knowledge. And then we believe, or have faith (if you like), in this knowledge. Of course, though our intuitions sometimes mislead us, it is false that they always do. Belief is not the same as knowledge because we can also believe that which is unlikely or uncertain, or even necessarily false. The practice of statistical hypotheses testing asks us to believe or have faith in the uncertain, in the unproved. The error is to assume that knowledge or probability and belief or faith are identical.

There is also a scurrilous definition of *faith* that it pleases some to state (see the *Skeptic's Dictionary*, [38]), which goes something like this: "Faith is believing *contrary* to evidence." It is possible to believe something you know is false, but the act is bound to cause distress. For example, I may claim to believe that I do not exist, based on who knows what evidence, but I am forced to confront myself when making the claim, which is psychically painful. I have to discount the knowledge, to pretend it doesn't exist while knowing it does. Doublethink. If I say, "I take it on faith that I don't exist", then this would fit the skeptic's definition. But nobody really believes statements like this. The proposition "*I* don't exist" starts with its own disproof.

Anyway, the kind of skeptic who says faith is believing contrary to evidence is substituting sloganeering for actual argument. He has a set of premises which lead him to knowledge or high certainty of some proposition, call it "not-X" (e.g., "God does not exist"), and he calls his premises "the evidence", which is fair enough. Except that his opponent has a different set of premises which he too calls "the evidence". Who is right? Well, he who can show a valid sound deduction, he who does not mistake a conditional for a necessary truth. My favorite sound valid deduction in this vein about this proposition is by [73].

A second tactic is for the skeptic to claim he has found a flaw in a proof for X. This may even be a genuine flaw for a given argument. If it is, and the skeptic is unable to persuade his opponent of it, but this opponent still claims to believe X based on the (flawed) proof, then the skeptic has a good example of somebody believing a claim contrary to evidence—but not contrary to faith. Love of bad arguments happens simply and frequently because most people are not well equipped to judge philosophical arguments at a deep level. What usually happens is that the opponent will hear the claim that a skeptic has found a flaw, and he might even believe this skeptic, but the opponent will still believe on other grounds. And this is not unreasonable unless the skeptic offers a necessary valid sound proof of not-X. If the skeptic hasn't, then he commits the fallacy of supposing that demonstrating one argument for X is flawed then all are. When this happens, what the skeptic really wishes is that everybody would be like him.

This digression is not as odd as it might seem. Arguments shooting past their targets are found everywhere. Scientism and the politicization of science have increased the kinds of fallacies noted here.

1.7 Belief and Knowledge

The word *belief* is ambiguous: statements of belief can belie knowledge, certainty, faith, or even uncertainty. You can only know what is true, but you can believe many things. Belief (the word) is often accompanied by the idea of lying; many people lie and say they believe a thing, while secretly doubting or disbelieving. This is what makes politics. The dependability of a person's public utterances accurately matching his actual state of mind depend strongly on his milieu. In repressive or totalitarian societies, like in the Soviet Union and some Western universities, the correspondence between public avowals and belief can be weak, or even negative.

We have to be careful and settle on one of the many definitions of *belief*. *True belief* (or just belief) is averring to or the acceptance of a conditional or a necessary truth. It is assent, or the acting as if some proposition were true, either necessarily or in the circumstances. As said above, belief is an *act*, a decision; it is not knowledge itself. I should believe conditional truths like "George wears a hat" given "All Martians wear hats and George is a Martian". I had *better* believe it. Why? Because the rules of truth and of logic demand it. If I doubted, which is to say if I did not believe "George wears a hat" given this evidence, it *must* be because I am using different evidence than the propositions "All Martians etc." What this different evidence is doesn't matter, but I *must* have it. I may *claim* to hold with "All Martians etc." but if I still don't believe "George wears a hat" then I must also be accepting other evidence which contradicts or trumps "All Martians etc."

We're finally ready to tackle *knowledge*, which is a necessary truth. You cannot have knowledge of a conditional truth, but you can believe one. Rather, you can have knowledge of the conditionality of conditional truth. Knowledge is sometimes called "justified true belief", the justification being that chain of sound valid argument which leads to indubitable axioms. This means (though we haven't yet discussed them) we can't have knowledge of probabilistic propositions (we can surely understand the propositions themselves, of course). It will turn out that propositions like "Given the evidence, the probability of X is p" (the entire thing inside the quotation marks) is itself necessarily true: p is not true, mind, but the proposition in which it appears is.

Succinctly: we only know and must belief necessary truths, and we cannot know but can believe (and usually do) conditional truths.

There are other (more confusing) ways to think about knowledge. Here I paraphrase the well-known ideas of Laurence Bonjour [24, pp. 27–52] (and use his notation p for a proposition). In order to know (the truth of) a proposition p in the "Cartesian conception of knowledge" (a theory!) three conditions must be met, the first two of which are: a person must believe or accept the proposition p without harboring doubt, and the person must have a reason or justification that guarantees the truth of p. The third condition is the strangest: p must be true.

But Bonjour, like most authors, does not separate necessary from conditional truths, nor do most authors recall the goal of the analysis of belief. I shall keep the distinctions. There are always two aspects to consider: whether something is necessarily or conditionally true, and what argument somebody is using to arrive at their proposition of interest. The failure to recognize these distinctions in truth opens up a curious situation called Gettier "problems".

Here is an example. In a standard raffle somebody must win; via the rules of such games we therefore know and believe that p = "Somebody must win." This is an existence proof, a statement of ontology, and a conditional truth. It is not a necessary truth because there is nothing proving it is logically necessary the raffle goes as planned (for instance, it may be played in Chicago). Who will win we do not learn until the drawing. If you are in the raffle it is therefore conditionally true, given the premises about standard raffles, that p = "I might win". You *believe* this given the accepted rules of raffles and because you own at least one ticket. The conditional truth of p is the reason and justification for *believing* p; it is also the proof p is conditionally true. Again, p is not necessarily true.

The example is worth giving because of so-called Gettier problems, named for Edmund Gettier the man who first inflicted them on philosophy [24, pp. 43–45]. Gettier claimed there were situations in which a person has a justified true belief, yet that belief does not meet the test of knowledge because the statements p are not true. Keep p = "I might win" which you believe is true because your wife said she bought you a ticket for the raffle. Yet your wife was teasing; she didn't buy a ticket, she only told you she did. However, unbeknownst to her or you, your mother did in fact buy you a ticket. Therefore you believe p, and indeed p is true, but, says Gettier followers, your belief cannot count as knowledge because your belief is based on a fiction (your wife's joke).

Naturally, I do not account situations like these as problems in understanding uncertainty. Since truth is conditional, the conditions you use to judge the truth of p—your wife said she bought a ticket, your wife told the truth, the rules of raffles, etc.—prove p conditionally. That is, given you accept those premises p is true, you should believe p. P is also conditionally true given the alternate premises "your wife lied and your mother bought you a ticket" (and removing "your wife told the truth"). P is also conditionally true if you live in Chicago, don't have a ticket but you get the wink from your alderman. There are many ways for p to be conditionally true. Your belief is driven by p's truth conditional on whatever evidence you used to prove p conditionally true.

But p is not necessarily true; you do not have knowledge that it is. No one does. There is therefore no problem with the concept of knowledge as justified true belief. Why? The outside observer who is aware of what both your wife and your mother has done, and who also is aware of the rules of raffles, also believes p is true, though in his case he is closer to a necessary truth because he has removed more of the contingency than you have. And once again, as must be repeatedly emphasized, you can still can believe your conditional truth and act on it; so can the outside observer who knows of your wife's shenanigans and your mother's beneficence.

Gettier "problems" stem from misunderstanding which evidence is being used at what stages and by whom to judge conditional and necessary truths. As long as you keep these clear and distinct, the "problems" disappear. To be clearer: you argue from the premises "My wife bought me a ticket and this is a standard raffle" which is probative of p = "I might win". P is true given the premises. But we *know*, i.e. we have justified true belief, that p is false given "The wife did not buy him a ticket." We need the JTB account of knowledge in order to argue that something is wrong with it! The goal here is the problem. Is the goal to ensure p is true? Or is it to ensure the premises are themselves true or clean from as much contigency as possible? If our goal is to make predictions of p's truth, then you will have made an accurate prediction. But if the goal is to assess the truth of the premises, then even though you are correct about p, you still fail because your premise, given the outside premise about the fact of the matter, is false. That means you only have *local* or conditional justified true belief because you only accepted contingent premises. But since most of the premises we accept are contingent, most accounts of JTB are contingent in the same way the lottery example was.

Because some consider the JTB account of knowledge to be "problematic" (again, how do they know this?) there have been many attempts at "restoring" the idea of knowledge to philosophy, such as "virtue" or "luck" epistemology; see [117] for many others. There is no hope of covering all these thrusts here, but it's worth examining very briefly the idea of "epistemic luck"; see [177, 178]. The idea is that, in the absence of JTB, an "agent" (by which philosophers always means a person but somehow can't bring themselves to say) hits upon an observable premise that is true. For some reason these accounts always focus on observables and not non-observable propositions that can be learned via induction (see Chap. 3). At any rate, suppose you are confronted by a multiple-choice quiz, with possible answers A–D. D, as it turns out, is the answer designed as correct by the teacher. But suppose you have no idea what the correct answer is, but you don't want to leave the answer blank, so you choose D. So you "win", just like in the lottery. Some philosophers want to say you hit upon knowledge because of your lucky guess—and it *was* luck. But this again mixes up the goal of the analysis. If your goal, as an epistemologist, is to check correct predictions, then indeed you nailed it. If the goal is instead to check the premises, you have failed. Why? We learn later that you must have been arguing from premises similar to, "There are four possibilities, only one of which is correct, and I must select one." The probability, given this premise, that you are right is 1/4. You might change the premises, if you know something of the subject and grammar, knowledge of the teacher, and so on; but none of that makes any difference. The point is that your premise was false in the sense that it *should* have been (something like) "Oho, I know what the question means and only one answer is possible." Whether anybody can get a student to admit that he didn't have the right premise and was only guessing limits the ability of a premise-analysis, which shows why multiple-choice questions are so poor at assessing knowledge. At any rate, luck had nothing to do with actual knowledge, which is always formed by premises of some kind. In this case, the premises were wrong. Justified-true-belief is in no danger.

Because *true* is such a useful word, and because *necessarily true* and *conditionally true* are cumbersome, like most people in ordinary speech, I will use *true* to mean either, relying on the context in most cases to define whether we have a conditional or necessary truth. But if there is sufficient ambiguity or the subject important, I'll spell it out.

Chapter 2
Logic

"All cats are creatures understanding French," said Alice's father. "And some chickens are cats."

"Wait, I know!" said Alice, chirruping. "That means that some chickens are creatures understanding French."

"What you said is true, my dear," said Alice's father, his voice full of pride.

What Alice said was true in the conditional sense that given or accepting or conditional on the evidence or premises or observations "All cats are creatures understanding French *and* some chickens are cats" then "Some chickens are creatures understanding French" logically follows. The conclusion is conditionally true.

Of course, Charles Dodgson knew, and we all know, that there are no *chats qui comprennent le français*, and that being so it cannot be true that *des poulets comprennent le français*. Which is to say we know these propositions are false. How? Because all evidence we have of our feline friends insists none understand French. Cats are diffidence personified and refuse familiarity with any language save their own.

So the proposition "Some chickens are creatures understanding French" is both true and false. There is no contradiction. It is true based on one set of premises, false on another. Logic says so; see *inter alia* [142]. Logic is the science or study of connections or relations *between* propositions, and to say an argument is true or false is to speak of the *relation* and *not* strictly of the propositions, thus when any proposition in an argument changes, the relation is liable to morph, too. The relation between Alice's evidence and the proposition is therefore true, and the relation between our observational evidence and the proposition is therefore false.

Dodgson, writing as Lewis Carroll, said his propositions were [36, p. 57]:

so related that, if the first two were true, the third would be true. (The first two are, as it happens, not strictly true in our planet. But there is nothing to hinder them from being true in some other planet, say Mars or Jupiter—in which case the third would also be true in that planet, and its inhabitants would probably engage chickens as nursery-governesses.

© Springer International Publishing Switzerland 2016
W. Briggs, *Uncertainty*, DOI 10.1007/978-3-319-39756-6_2

They would thus secure a singular contingent privilege, unknown in England, namely, that
they would be able, at any time when provisions ran short, to utilise the nursery-governess
for the nursery-dinner!)

Since probability, the main focus of this book, is the continuation, fullness, or
completion of logic, and since logic is the study of relations between propositions,
therefore probability is also the study of relations between propositions. Note
carefully: *between*. With that conclusion we are done: we have everything we need
to know; this is the complete "theory" of probability and statistics. The rest is mere
detail.

2.1 Language

Aristotle (again) says our knowledge of truth begins in sense experience. But not
everything we know or is true can be sensed, except in the weakest form of that term
by which "sensed" means the workings of our thought process, which can be felt
not as muscle movement or nerve excitement, but as mental images and exertions
and so forth. There are three acts of the mind: apprehension, judgment, reasoning.
We need to understand each, at least broadly.

Apprehension is learning the content of each argument. We first need to
apprehend the nature of each word and also the grammar in which propositions
are written. Ambiguities are more than possible, especially when asking questions.
For example, how happy are you? On a scale of one to eleven-point-four, of course,
in units of the seventh root of π (numbers make this science). There are many who
find this question comprehensible. I do not. Happiness we can grasp, but arbitrarily
indexing it to a number just so it can be manipulated by well-loved equations I do
not follow. This question and its multitudinous cousins are responsible for a great
deal of scientism and over-confidence. About these subjects, and about ambiguity,
more later.

Every term, every universal, has extension. *Tree* is a term, individual trees are
its extension. *Happy* is a term, individual instances, mental states of persons, are
its extension. Every term also has intension (not inten*t*ion) or comprehension.
Intensions are those attributes or qualities that make up the notion of the term.
These are important probabilistically. We have a way of speaking which indicates
universality which does not follow strict rules for syllogisms, but which nevertheless
conveys truth. For example, when we say "Men are taller than women" we do not
imply that the shortest man is taller than the tallest women. Instead we mean it
is in the *nature* or *essence* of men to be taller than women, a truth conditional
on extensive observation and the induction to the generalization. Men taller than
women is what we *expect* to find (I use this word in its plain English connotation).
This is also to speak probabilistically: the sentence implies there is some high but
unquantified chance that any given man is taller than any given woman. Stereotypes
also follow these rules. Steven Goldberg notes that most stereotypes are true in this

probabilistic sense, but that people's conception of *why* they are true is often at fault, [97, 98]. It is the *why* that is a major concern to us in this book.

The second act of the mind is judgment. Our concern is with propositions, sentences which can be true or false or somewhere in between. In logic we move from one, or one set, of evidentiary propositions to a conclusion, which is another proposition. In strict logic there is the idea that the conclusion is derived or deduced, and while this happens in formal cases, in reality it is we who specify the evidentiary propositions, also called premises or evidence or data, and we who specify the conclusion or proposition of interest. This freedom is what gives rise to the fallacy of "subjective" probability, as we'll see.

The last act of the mind is reasoning, the activity that separates men from brutes. For instance, we *reason* (here, a verb) that "P is not not-P" and that tautologies are truths. Tautologies have a special place in logic and in probability. Examples: "If it is raining, it is raining", "You either have cancer or you don't", "All gloomy people are gloomy", or the old classic "All bachelors are unmarried men." These propositions are all necessarily true, *given* our innate knowledge of logical rules and of the words themselves, given, that is, our understanding of the nature of logic and the intension of the terms. Since tautologies are necessarily true, they add nothing to any argument. What good and what insight, after all, is there in telling a woman, "You either have cancer or you don't"? None, of course. And nothing is added if you change the tautology to "You either have breast cancer or you don't." Yet that tautology is suddenly seen full of fearful importance. (Switch "breast" with "prostate" for men.) And that is because people understand the grammar to mean more than it strictly implies. Words matter.

2.2   Logic Is Not Empirical

Much of this section is a paraphrase from David Stove's *magnum opus The Rationality of Induction*, a neglected, or rather unknown, work in probability, [206].

We cannot know all logical truths empirically. For example, there is no way to learn through observation the validity of the argument "'James is a man and Harry is a man' entails 'James is a man.'" We can of course observe the maleness of either individual, but we cannot witness the *entailment*, that which makes the argument true. Neither can we observe that "'X is a man and Y is a man' entails 'X is a man'", because witnessing each and every X and Y is impossible. Neither is it true that "'James is a man and Harry is a man' entails 'James is a man'" *because* "'X is a man and Y is a man' entails 'X is a man'"; it is true all on its own and not *because* it is part of some schema or formal theory. It is not our organizations of logic that makes true statements true: they are true on their own merits.

A matter of supreme importance can be teased from this. Here is a proof that we must come by knowledge that cannot be acquired solely by experience. The knowledge alluded to here are the rules of logic, the very steps in reasoning, the *how* we know when something is true or false.

This example is also from Stove (modified slightly). In order to know via experience the validity of (say) the schema A = "For all x, all F, all G, either 'x is F and all F are G' is false, or 'x is G' is true", we could make observations like O_1 = "David is bald and David is a person now in this room and all persons in this room are bald." But in order to get from O_1 to A; that is, to know A is necessarily true, we have to already know that O_2 = "O_1 confirms A", and that is to have non-empirical logical knowledge. Or you could insist that O_2 was learned by experience, but that would require knowing some other logical knowledge, call it O_3, which somehow confirms O_2. And then there would have to be some O_4 which somehow confirms O_3, and so on. There cannot be an infinite regress—the series *must* stop somewhere, at a point where we just know (my guess is O_2)—so we must, are forced, to rely on induction (which is examined next chapter) to supply us things like O_2.

This isn't all. We can learn from observation the following argument is invalid: "'All men are mortal and David is mortal' therefore 'David is a man'" if perchance we see David is not a man (maybe he's a puppy). And we can learn from observation the invalidity of "'All men are mortal and Peter is mortal' therefore 'Peter is a man'" only if we see Peter is not a man (maybe he's a cow). But we cannot learn the invalidity of "'All men are mortal and X is mortal' therefore 'X is a man'" through observation because we would have to measure every imaginable X, and that's not possible. If we believe "'All men are mortal and X is mortal' therefore 'X is a man'" is unsound, and it surely is, this belief can be informed by experience but it cannot be solely because of it that we have knowledge of it.

Stove: "If an argument from P to Q is invalid, then its invalidity can be learnt from experience if, but also only if, P is true and Q is false in fact, and the conjunction P-and-not-Q, as well as being true, is observational. This has the consequence, first, that only singular judgements of invalidity can be learnt from experience; and second, that very few even of them can be so learnt." And here's the kicker: "If the premise P should happen to be false; or the conclusion Q should be true; or if the conjunction P-and-not-Q is not observational but entails some metaphysical proposition, or some scientific-theoretical one, or even a mere universal contingent like 'All men are mortal': then it will *not* be possible to learn, by experience, the invalidity of *even this particular argument*" (pp. 155–156). The key is that "scarcely any of the vast fund of knowledge of invalidity which every normal human being possesses can have been acquired from experience."

Examples? The invalidity of the argument "Given 'The moon is made of cheese' therefore 'Cats do not understand French'" cannot be learned from experience. Neither can "Given 'Men can breathe underwater unaided' therefore 'The atmosphere is largely transparent to sunlight'". In neither can we can ever observe the conjunct P-and-not-Q. Yet we know these are false. Why? Induction again.

We often in mathematics invoke the continuum, the infinity of numbers on the "real line", or of different kinds of infinities. None of these are ever observed and can never be observed, yet no mathematician doubts their truth. These and various other "puzzles" are solved by induction, a highly misunderstood concept, as we'll see.

2.3 Syllogistic Logic

Mathematical propositions are highly formal. It would not do, for example, to claim this mathematical equation is universally true: "$AB = BA$." The proposition is sometimes true, for instance when A and B are finite natural numbers, but it is in general false when A and B are matrices. Mathematicians in their proofs are thus sticklers for detail. Limitations and constraints on their propositions are laid out with excruciating rigor. Indeed, *rigor* is a high compliment. Sometimes these efforts are beautiful; sometimes they are ponderous. Only rarely does somebody catch out an error in a proof, and when it happens it is usually not because of a miscalculation but because a constraint not thought necessary actually was. As a result of this vigorous scrutiny, people trust mathematicians when they say something is so. But we must never forget that their proofs are true in relation to the premises used. And those premises are true only because of earlier premises in the chain of the proof, and so on down to the axioms which everybody believes true conditional on their intuitions (induction). This is what makes for a necessary truth.

Besides the usual mathematical expressions found in analysis, number theory, and the like, there is a formalization of logic which has various names like symbolic or mathematical logic and propositional or sentential calculus. These fields belong properly to mathematics, though they provide useful results to syllogistic logic, which is our main interest. Syllogistic or Socratic logic is meta or street logic, arguments as they are used in life and in science and statistics or in assessing the value of the more formalized logics. It is always there, the bulwark of everything else. Mathematical logics are no different than other mathematical subjects: proofs are given with meticulous assiduity paid to constraints on the symbols used, indeed to the very languages used, languages which (oft times) resemble actual speech not at all. But since we need ordinary words to have real arguments, we need to grasp the limitations, fallibilities, and the ultimate strengths of Socratic logic.

Syllogistic, two-valued, Aristotelian, plain-words logic is employed when philosophers attempt to prove the superiority of other logics or in describing the usefulness and necessity of mathematical logics, and even to explain why syllogistic logic is not to be preferred, see [164]. Even in *Principia Mathematica*, the book which taught us (eventually) $1 + 1 = 2$, Alfred North Whitehead and Bertrand Russell were obliged to use plain language to describe what their symbols meant.

Language and not mathematics is the tool we're stuck with and which we must use to express ideas, such as certainty and uncertainty. If you disagree, write me a letter stating why. Syllogistic logic is written in ordinary language, which is always and necessarily found at the start and finish of any argument or analysis, including scientific analyses. It's there when we tell our audience what the results mean and what we should do about them. Syllogistic or meta-logic is needed in mathematics, too, especially in the branch known as applied mathematics. This is when mathematical, which is to say purely metaphysical, ideas are applied to real life contingent, i.e. physical, processes.

Every time an equation is called in to support, say, how much weight a bridge can hold, syllogistic logic comes in play. The equations used in support of engineering have no meaning by themselves; they must be given meaning by us. The arguments which support these labelings are difficult and drawn out, but they are all examples of syllogistic logic. Since these arguments involve physical principles, i.e. contingent events, the end results are always at best conditional truths, and sometimes only probable truths. And occasionally even nonsense. I have seen equations applied to human behavior, usually in economics or social science, all mathematically spotless, which when applied to real people are gibberish. The problem is that people commit the simple fallacy, "Since the equations are mathematical truths, the objects to which they are said to apply must be those objects, and therefore the predictions and theories which gave rise to the mathematics are therefore true." We'll meet these fallacies later when we discuss models.

Besides, if we opt for symbolic logic we're apt to take perfectly understandable propositions like this "Socrates is wise" and turn them into curiosities like this: $(\exists x)(Sx \& (y)(Sy \leftrightarrow y = w) \& Wx)$, an example from [164, p. 191]. As useful as this sort of thing is to understand the fine shadings of mathematical logic, it is a positive bar to clear understanding of real problems.

2.4 Syllogisms

There isn't much point rehearsing the kinds of syllogisms, enthymemes, major and minor premises, *barbara*, *celarent*, and other staples of logic. These are all too well known, and there are many texts which do a superior job, see [134]. Even high schoolers still know that given "No academics have a sense of humor and all teachers are academics" that it is conditionally true that "No teachers have a sense of humor." No: what we have to understand is what *kind* of truths syllogisms give us.

As the last example showed, syllogisms can give us conditional truths. Since the premise—written as one single premise, cobbling the major and minor premise together, but there is no difficulty in doing so—is known not to be true in your author's case. The proposition "No teachers have a sense of humor" is, with respect to that evidence, universally false. Thus it is false and true simultaneously, depending on which set of premises one chooses to believe or employ. But don't forget, any set of premises includes tacit knowledge of the meaning of the words used. This is inescapable.

Some syllogisms give universal truths. Given "All men are mortal and you, the reader, are among the race of mankind" then the proposition "You are mortal" is necessarily true because the premises are known, through a chain of sound observation and argument, to be necessarily true. Don't wait forever to make out your will.

As said in Chap. 1, in ordinary speech we'll say the conclusions of both arguments are "true", which can be harmless because most people take the point. But since language can be ambiguous, we have to take care to say just we mean when speaking formally, or when discussing sensitive topics like politics. An example in line with our ultimate goal: given "Some systems of government are stultifying and all stultifying systems of government are deadly" then it is true that "Some systems of government are deadly." The proposition is therefore true and not probable, though it carries the sound of probability because of that "some".

Later we will learn that given "The rules of logic and mathematics and accepting this and such evidence which is probative of Y" that the propositions "It is probable that Y", "Y has a $p\%$ chance", "Y might be true" and the like are themselves true. But conditionally true because although the rules of logic and mathematics are necessarily true, the evidence probative of Y will not be. It is we who specify that evidence, picking from an veritable infinite universe of evidence.

For this reason many, but not all, probability statements are conditional truths. An example: X = "Given certain evidence, the chance of Y is $p\%$". X itself, assuming the tacit premise of no miscalculation or other mistake in applying the evidence to Y, is necessarily true. But that the "Chance of Y is $p\%$" is only conditionally true based on the certain evidence.

One logical tidbit which is awfully useful: Any valid argument from P to Q is unchanged if a necessary truth is added to the list of premises P. That is, P to Q is identical in truth value with P & T to Q, where T is a necessary truth (this is like multiplying an algebraic equation by 1). This applies to syllogisms and probability arguments. Another: it is impossible to deduce necessarily true consequences from contingent premises. Believing the opposite is like trying to support the earth with turtles.

2.5   Informality

Because street logic is informal it is not possible to constrain the reach and type of propositions used. Anything goes. This freedom, as all freedoms do, comes at a price. Each argument must be judged on its merits; judged individually, I mean. Truth tables, proof by parallel argument, similarity to a set of symbols said to represent arguments of this or that schema, and the like are therefore not useless but are of limited applicability.

Stove [206] defines formality in an argument as when "it employs at least one individual variable, or predicate variable, or propositional variable, and places no restriction on the values that that variable can take". Stove claims that "few or no such things" can be found. This will be useful for us to recall when discussing the hideously complex regression models that are much in fashion in some circles. The so-called rule of transposition is an example of what formality in logic might look like. The rule is: the proposition "If p then q" entails "If not-q then not-p" for all p and for all q. This is formal in the sense that we have the propositional variables p and q for which we can substitute actual instances, but for which there are no restrictions. If Stove is right, then we should be able to find an example of formal transposition that fails.

First a common example that works: let p = "There is food" and q = "I can eat", then "If p then q" translated is "If there is food I can eat". By transposition, not-q = "I can't eat" and not-p = "There is no food" thus "If not-q then not-p" translates to "I can't eat if there is no food." For an example in which formal transposition fails (this is Stove's, too), let p = "Baby cries" and q = "We beat him", thus "If p then q" translates to "If Baby cries then we beat him". Heartless; but logic is a hard taskmaster. But then by transposition, not-q = "We do not beat Baby", not-p = "He does not cry", thus "If not-q then not-p" translates to "If we do not beat Baby then he does not cry." And this is obviously false.

So we have found an instance of formal transposition that fails, which means logic cannot be "formal" in Stove's sense (I do not intend to give a full proof of this here). It also means that theorems which use transposition in their proofs will have instances in which those theorems are false if restrictions are not placed on its variables—it's the restrictions that are important. It's actually worse than this, because transposition is logically equivalent to several other logical rules, putting those theorems in jeopardy. Those who prove theorems are, however, usually very careful detailing restrictions and, as said, those theorems found to have failed usually suffered from lack of complete or improper restrictions.

It is Stove's contention that all logical forms have examples where the logic is turned on its head, like with transposition, *unless*, like in formal mathematics and mathematical logic, restrictions are in place. As said above, it is not a universal truth that "$AB = BA$." But it is when we add the restriction to (say) natural numbers. This means we have to be very careful in saying what are the precise conditions and limitations of our models.

2.6   Fallacy

Not all fallacies are what they seem. Given "All dogs have four legs and Iris has four legs" it does not follow that "Iris is a dog", not because in some formula or schema like "'X is F' does not follow from 'All F are G and X is G'", but because it might be that Iris is a cow or some other creature with four legs. It is because we can summon evidence about the range of these alternate possibilities that the truth of the proposition is in question. Symbols in formulae and so forth are scratches on a page and take no meaning until we supply one, thus symbols or schema can't be true or false. Only arguments can.

To avoid fallacy, we always must take the information or evidence supplied as given and concrete, as sacrosanct, even, just as we do in any mathematical problem. Therefore we accept *arguendo* that "All dogs have four legs", even though by our observation we know some do not (never play fetch near a highway). We know of the four-leggedness of dogs because that is their nature, just as we know three-legged dogs are deficient or are suffering from a privation.

It is not true that "Iris is a dog" given "All dogs etc." But that does not imply the logical negation of our proposition, i.e. it might be so that Iris is, in fact, a dog. Another way of putting it is that given "All dogs etc." it is not false that "Iris is a dog". The status of our proposition, conditional on the only evidence we have, is murky. Relying solely on dichotomous logic leaves us hanging. The logical status of "Iris is a dog" is not empty, though, and it would be a mistake to think it was. Here we know Iris has four legs; dogs do, too. Iris therefore *might* be a dog. We just don't know for certain she is. The ambiguity drives us to probability, where we complete our understanding of the proposition. There is no fallacy in the argument *unless* somebody insists or implies it is *certain* Iris is a dog.

A depressing but far from unknown fallacy goes: "If X then not-P, P, therefore not-P." The variable X is a cherished theory or belief, usually a popular or faddish theory of human behavior, but also perhaps a physical theory which has powerful interests, and P any proposition about that behavior or a state the theory says cannot happen. After P is observed to occur, the theory X would appear to be in difficulty. But human ardency is infinitely malleable, especially if X is the creation of its believer. P is denied, or perhaps the No True Scotsman is invoked for P, or an R is invented such that the original argument is modified to "If X-and-R then P or not-P, R, etc.", where R is an excuse; anyway, X survives. This will be important when we discuss falsifying models.

Another popular fallacy in studies which use statistics is the *peer-review* or *credential* fallacy, which appears in several forms, as variants of the *Appeal to Authority* and *Genetic* fallacies. The most common is in journal writing, where an author will write, "Jones (1999) showed that X", with the implication (at least sometimes) that X is therefore true *because* "Jones (1999)". The reference is offered as sufficient proof, especially if the journal in which Jones's work appeared is prestigious. Usually authors will clump a dozen or so references as a useful summary, and this can be a move to bludgeon the reader into submission. If several authors write on a doubtful proposition, the mass of citations is often taken as proof. Hence intellectual fads have strong inertia in our publishing age. Physicians also sign their names on papers as "John Smith, *M.D.*" for the same reason. Members of the public, though very often academics, too, especially on political subjects, will refuse to listen to an argument of an opponent unless it first be ensconced in a "peer-reviewed" publication. These are obvious and perennial fallacies, but still unfortunately persuasive. Since they have been with us forever, it is rational to conclude (via induction, which we discuss next chapter) that they always will be.

There are many other fallacies, which will be dealt with in turn, when they are more specifically tied to certain statistical procedures, such as the *epidemiologist fallacy*. Formal fallacies, broken syllogisms and the like, are easy to spot, and when they are a necessary truth has been discovered, which is the complement of the fallacy. Formal fallacies aren't especially rare, either, and are found in increasing frequency the further one gets from mathematics. Particle physicists, say, generate few formal fallacies, but literature and social science professors are positively bursting with them. We'll examine the more common of these in due course.

Perhaps the worst fallacy is the We-Have-To-Do-Something fallacy. Interest centers around some proposition Y, about which little is known. The suggestion will arise that some inferior, or itself fallacious, method be used to judge Y *because* action on Y must take place. The inferiority of fallaciousness of the method used to judge Y will then be forgotten. You may think this rare. It isn't. An entire branch of statistics, hypothesis testing, is built around this fallacy. We come to this in time.

Chapter 3
Induction and Intellection

"[W]hen Mr Wells says (as he did somewhere), 'All chairs are quite different,' he utters not merely a misstatement, but a contradiction in terms. If all chairs were quite different, you could not call them 'all chairs.'"—G.K. Chesterton

There is no knowledge more certain than that provided by induction. Without induction, no argument could, as they say, get off the ground floor; this is because induction *provides* that ground floor. No argument could even be phrased if it were not for induction, because phrasing requires language and language requires induction. When we say *apple*, we know it applies to all apples.

All arguments must trace eventually back to some foundation. This foundational knowledge is first present in the senses. Through *noesis* or *intellection*, i.e. induction, first principles, universals, and essences are discovered. Induction is what accounts for our being certain, after observing only a finite number of instances or even one and sometimes even none, that all flames are hot, that all men are mortal, that *white* is always white, that for all natural numbers x and y, if $x = y$, then $y = x$, and for the content and characteristics of all other universals and axioms. Because we know these indubitable propositions more surely than any other, induction produces greater certainty than deduction.

Arguments are not valid because of their form. It is not because of *barbara* (a common syllogistic form) that because all men are mortal and Socrates is a man that we deduce Socrates is mortal. That conclusion is made obvious to us by observing the congruence of all the propositions in the argument, and it is induction that tells us the major premise is true and it tells us that the congruence *provides* or *gives* proof. But induction only works for telling us *what* we know and not *why* a necessary truth is *caused* to be true. Indeed, the standard story is backward: *barbara* and other syllogistic forms are discovered from instances like "All men..." via induction. Induction provides the *certainty* that, given these premises, the proposition or conclusion is true. Just as it provides the certainty that the probability of "George wears a hat" is 50% given "Half of all Martians wear hats and George is a Martian".

Mistakes in induction occur, as they do in every area of intellectual activity. When a man sees several white swans and reasons, "All swans are white", he is proved wrong when a black swan in sighted (as in Australia). Why he is wrong is explored below. When a poll, asked of one set of people, is enlisted to "work" on a second set of folks who differ in those characteristics responsible for first set's answers on the poll, again induction fails. But statistical induction, of which polls are instances, is only one kind of induction. Induction is of five different kinds, some more and some less strong.

Since at least Hume it has been fashionable to pretend mystification about why induction is "justified" or to claim that it is not. Hume said, "We have no reason to believe any proposition about the unobserved *even after* experience!" [116]. Howson and Urbach [115], for instance, (p. 4) say that there is no "solution" to induction and that this sad fate "is no longer controversial." Speaking of induction, Karl Popper asked, "Are we rationally justified in reasoning from repeated instances of which we have experience [like the hot flames] to instances of which we have had no experience [this flame]?" His answer: "No" [175]. He also said, "The truth of any scientific theory is exactly as improbable, both *a priori* and in relation to any possible evidence, as the truth of a self-contradictory proposition" (i.e. It is impossible; see also [176]).

Fisher, though not of the same skeptical bent—he often talked about how scientists used inductive reasoning, though he wasn't always entirely clear by what he meant by "inductive" [77, 78]—agreed in principle with Popperian ideas and used these beliefs to build his system of statistics. Theories (propositions) could only be "rejected" and never verified and so on. Popperian skepticism is common in statistics. For example, a well known statistician on his blog wrote "[I]nduction doesn't fit my understanding of scientific (or social scientific) inference" [92]; also see [93] for the standard limited view of induction in statistics. As we will see, such skepticism is unwarranted.

3.1 Metaphysics

Why a section on metaphysics in a book about probability? Because probability, like the philosophy of anything else, must rest on some ground. That ground is our knowledge and understanding of essence, substantial forms, intellection, on the quiddity (the *whatness* of things) and (in the old way of speaking) induction.

Material things are composed of matter (or matter and, equivalently, energy) and form, and the study of such is called *hylemorphism*. A simplistic example: the same lump of clay may be fashioned into a vase, an ashtray, a fanciful backscratcher, or many other things, each of a different form. A substantial form of an ashtray would be those elements, or rather arrangements of matter, that make the clay an ashtray and not, say, a backscratcher. We might say an ashtray has a bottom to collect ash, and so a sculpture of a donut made of clay would lack the substantial form of an ashtray. An accidental form is something part of the substantial form which is not essential. For instance, the ashtray can be an ashtray with a carved initials or without. The initials are a form, but their absence would not remove the substantial form of the ashtray.

The *essence* of a thing is its nature, its whatness. Everybody knows that a chair is not a house nor is water mercury nor are men chickens, except metaphorically. But even metaphors need induction to work. Everybody, even non-scientists, knows there are differences between kinds of things because they understand the essence of different objects—chairs, houses, water, mercury, men, chickens—and they know at least partially the reasons behind that which makes them what they are, even if they do not comprehend the totality of their essences. Nor need any man understand all there is to know of a thing to understand part of its essence. That things have natures and the study of their substantial forms is called *essentialism*. The best book for readers of this volume is David Oderberg's *Real Essentialism* [166], where these matters are defined and defended at length and (I think) conclusively.

All scientists and all users of models of uncertainty take essentialism for granted. Whenever "data" are collected, they are collected on *like* objects, objects which have the same essence (or are thought to). A doctor testing a new pill collects measurements on people, and not people, ferns, and Buicks. A physicist measuring heat ensures to the best of his ability that the apparatus and experimental milieu remain the same or similar for each iteration. The essence of interest for the doctor, though this is usually unacknowledged, is human beings. He knows what is a human being and what isn't. And similarly for any controlled experiment: essences are a given. This is not to say that mistakes in identifying essences aren't made. Finger bowls are drunk from. But there can't be accidents like these without essences (this will turn out to be a good pun). Science is (a weaker pun) the essence of discovering essences.

Deduction assumes essence. If we accept "All men are mortal" and "Socrates is a man", it is deduced that "Socrates is mortal" *because* we know that Socrates, being a man, shares in the essence of men. There is more to arguments than that, of course, because we also have to understand *mortal*, *all*, *is*, and so forth. All of these words, some of which describe essences and some other things, are known inductively.

To expand on the examples above, the essence of thing is not merely a collection of its properties. The lumberyard, which contains all the necessary components for a house, is not a house. We need to marry the material of the house with its substantial form for the house to come to be. An *accident* is a property that does not necessarily have to belong to an object for it to retain its essence. Houses can be white or green; a house's color is an accident; that is has a color is part of its essence. Humans can wear shorts, dresses, pants, or nothing; clothing is an accident. Everybody can grasp that this is a house or that is a human regardless of the accident of color or clothing. It is a necessary or *essential property* of houses to have roofs and for humans to have hearts, even "artificial" ones (the word *acknowledges* essence). It is the essential properties of a thing that define the thing. A roof-less house would not be a house but a shell without a cover; likewise a human without a heart is not human but a corpse. Suppose a house is under construction and lacks a roof because of a hurricane. That some house-like objects do not have roofs does not imply that houses don't have to have roofs to be houses, or that the essence of a house is impossible to define. A three-legged dog which has ignored the advice about playing in traffic is not a stool. Everybody knows that four legs are part of the essence of being a dog. Scientists who study dogs and those who breed them know more about what is essentially a dog than most people, but again the entirety of an essence does not have to be known. People can tell the difference between dogs and cats and kumquats. Some dogs are missing legs and some are missing ears, just as some are black and others brown. These are all accidents. We come to know the essence or nature of a thing by recognizing its essential properties and not by its accidents.

This too is acknowledged by the experimenter. The doctor trying out a new prostate treatment gives it to many men. Each of these men have the same essence—they are men!—but they differ in accidental ways. Some are taller, some shorter, some have gene variant G_1 others variant G_2, some grew up in this region and others that, some smoke and others don't, and so on. It is not that accidents are ignorable. In the doctor's experiment, many accidents to being a man will be measured because it is thought or conjectured that the accidents, or some combination of them, interact differently with the causal powers of the treatment. I take up this again in the chapter on Causality, but it should be obvious that if there were no accidents between (or in) the men and the experimental protocol was identical for each, then the outcome would be of the same nature for each; where the "same nature" might not mean "identical" but could, but where it does mean "in a known stated range". This will make more sense when discussing causality and quantum mechanical events. If there were no accidents but the protocol varied between men, however slightly, then outcomes could be different.

3.2 Types of Induction

One reason induction is widely misunderstood, even considered a "problem" in the academic sense, is because it is analogical. Mistaking one use of induction for another is equivocation, and, as I stated, equivocation makes the best jokes[1] but the worst fallacies. As with all analogical terms, sometimes it is harmless to leave a word unadorned and sometimes it is not. In this section, and elsewhere when needed, I hyphenate the word to indicate which type of induction is meant.

In this chapter I largely follow Louis Groarke's wonderful *An Aristotelian Account of Induction* [102], which is must reading, especially given the controversy over this topic. Also see [103]. Groarke's work is largely unknown in probability and statistics, but it shouldn't be. There is no way to adequately summarize the entire work, which is long and deep. Only a few highlights sufficient to dispel the sense that induction is problematic are given here.

"The goal of induction," Groarke tells us, "is not simply to *prove* that something is the case but to provoke an understanding of the general case." We here and elsewhere follow the dictum *nihil in intellectu nisi prius in sensu*, "there is nothing

---

[1]Two cannibals are eating a clown and one says to the other, "Does this taste funny to you?"

in the intellect which is not first in the senses." Our senses tells us what is the case. Induction moves from the particularities collected by the senses, and moves to unobservable, unsensible generalities or universals, such as knowledge of a thing's essence. Induction starts with the finite and progresses to the infinite; so although we can never entirely grasp the infinite, we can and even must know part of it. Induction, Aristotle's *epagoge*, comes in different flavors; at least five. There is no appreciation of this that I have ever seen in the uncertainty, probability, or statistics literature, except in the works of David Stove, David Williams, and a few others. But, as we shall see, even they only "justified"—by which I mean clarified what was already true: no human can "justify" any necessary truth—one form of induction, i.e. induction-probability, which most people already believed unproblematic. Certainly all probabilists and statisticians *acted* as if induction-probability were useful.

According to Groarke's view, induction is "the cognitive/psychological mechanism that produces the leap of insight" necessary for all understanding. He gives five flavors, aspects, or facets of induction. These are (in my modified terms more useful for probability) (1) *induction-intellection*, (2) *induction-intuition*, (3) *induction-argument*, (4) *induction-analogy*, and (5) the most familiar *induction-probability*. The order is from that which provides the most to the least certainty.

Induction-intellection is "induction proper" or "strict induction". It is that which takes "data" from our limited, finite senses and provides "the most basic principles of reason." Senses provide information of the here-and-now (or there-and-then), but induction-intellection tells us what is always true everywhere and everywhen. We move *with certainty* from the particular to the general, from the finite to the infinite. Without this kind of induction, no argument can ever get anywhere, no argument can ever even start; without it language would not be possible. Induction-intellection "Operates through infallible exercise of [*nous*], through the activity of intellection, understanding, comprehension, insight." It produces "Abstraction of necessary concepts, definitions, essences, necessary attributes, first principles, natural facts, moral principles." In this way, induction is a *superior* form of reason than mere deduction, which is something almost mechanical, and can be done on a *mindless* computer. Induction-intellection is *instantaneous* learning, it proceeds by "flashes" of insight. (*How* this happens is not a question here answered; *that* it happens is indubitable.) Intellection-inductions are not found in the slogging labor of mechanically working out consequences of accepted premises, like deductive reasoning is. The knowledge provided by induction-intellection comes complete and cannot be deduced; it is the surest knowledge we have. Numbers come from this form of induction. We see one apple, two apples, three. And then comes 1, 2, 3, ... Deduction has much to say about that "...", but knowing that we *can* reason deductively comes from this form of induction; see [81]. See [103] for a discussion of how induction works (or might work), a topic I do not here broach.

Induction-intuition is similar to induction-intellection. It "operates through cleverness, a general power of discernment of shrewdness" and provides knowledge of "any likeness or similitude, the general notion of belonging to a class, any discernment of sameness or unity." Axioms arise from this form of induction. Axioms are of course the ground of all mathematical reasoning. We have to be careful because some use the word *axiom* loosely, and merely mean *assumption*, a proposition which is not necessarily believed but is desirable: thus, by *axiom* I mean those base propositions which are fundamental and believed by all those who consider them (like Peano's axioms, etc.). This fits with work like Göedel's, who proved, as it were, that axiomatizing never stops, that induction-intuition must always be present, that not all things can be proved by reason alone. We don't know that syllogisms, for instance, provide validity except first by knowing via induction-intuition that syllogisms are deductive. The foundational rules of logic are provided to us by this form of induction. These rules do not occur to all—not everybody comprehends every truth, as is clear in any prolonged contact with students—but that rules *are* grasped (if they are) is done via induction and not reason. We observe that our mom is now in this room and now she isn't, and from that *induce* the principle of non-contradiction, which cannot be proven any other way. If mom is here, she can't not be here. Therefore, induction says, if a thing exists it can't simultaneously not-exist (our knowledge about thing's existence can be indefinite, of course). No universal can be known except inductively because nobody can ever sense every thing. Language exists, and works, because induction-intuition.

Induction-argument, given by inductive syllogisms, is the "most rigorous form of inductive inference" and provides knowledge of "Essential or necessary properties or principles (including moral knowledge)". The forecaster on television announces E = "It's probably going to rain tomorrow" and we're interested in the conclusion Y = "It will rain tomorrow". Y does not follow validly from E, but can can induce Y given the tacit premise P = "All the times I heard this guy on TV say it will rain he's been right". Y is now *certain* if we accept this additional premise. But that is a weak example. A better is when a physicist declines to perform an experiment on electron number 2 because he has already performed the experiment on electron number 1, *and* he claims all electrons are identical. Induction-argument can provide conditional certainty, i.e. conditional truth.

Induction-analogy is the least rigorous but most familiar (in daily life) form of induction and provides knowledge of "What is plausible, contingent or accidental; knowledge relating to convention, human affairs." This form of induction explains lawyer jokes.[2] Stereotypes fall under this form. As Goldberg has repeatedly shown us, [97, 98], stereotypes are often accurate, but the causes of these stereotypes is just as often in error. See also the work of Jussim on the correctness of many stereotypes [129, 130].

Induction-probability of course is the subject of most of this book. It provides knowledge of "Accidental features, frequency of properties, correlations in populations" and the like. It is, as is well known by anybody reading these words, the most prone to error. But the error usually comes not in failing to see correlations and confusing accidental properties with essences, but in misascribing causes, in

---

[2]What's the difference between a good lawyer and a bad lawyer? A bad lawyer makes your case drag on for years. A good lawyer makes it last even longer.

mistaking correlation for causation even though everybody knows the standard admonition. Knowledge of this "law" never seems to stop everybody violating it, though.

In trying to solve the "problem" of induction Stove and Williams separately showed that reasonably sized (but still small) samples can "resemble" their populations in essence. Their results should come as no surprise to any statistician familiar with sampling, though some philosophers over-fond of Hume and his progeny were taken aback: many do not seem to be able to resist the allure of doubt. On the other hand, Stove and Williams and their sympathizers do not appear to understand that they are demonstrating the soundness of only one type of induction, i.e. induction-probability.

In general, and across all types or flavors, an induction is an inference about the unobserved conditioned on that which has already been observed, where the propositions of both (unobserved and possibly unobservable and the observed) are in some way similar, i.e. share an essence. Induction is another way to speak of essences, with varying degrees of certainty depending on the type. But there is another sense of the word initiated by Carnap and others. To them, induction is any (believed) inference not deductively valid, which is a much wider class of inferences than what I mean by inductive inferences, an enormous class which contains even absurdities. "If the moon is made of cheese then it's noon" is "inductive" in this sense, thus it is not surprising that some see a problem with "induction" when it is put into such loose terms. Part of the difficulty is that modern logicians will call this proposition true because of its (logical) form whereas the civilian, scientist, and old-school logician will think the modernist is nuts. Why insist on such bizarre arguments! It's not form that is important, but cause and essence, knowledge of which is provided by real induction. Since our goal is to understand (logical) probability, I adopt the older meanings of induction, as given above. I also do not use, recommend, or like Carnap's use of the term "inductive probability". Probability is probability, but Carnap often used his term to apply to non-inductive arguments.

All these forms of induction have "stages" or levels. The first is ordinary, the simple noticing of how things are, like sugar tends to be sweet, ice floats, that people walk. The second is the identification of accidents, like all people on this street are wearing (or rather affecting) peasant clothing (jeans). The third is the abstraction of necessary properties, that people have two legs, that they breathe air, that one plus one must equal two. Finally comes the realization of essences, that having two legs, a heart, and the powers of rationality is what *makes* a human. Probability models, as we shall see, come into play in each.

Here is an example tying induction to essence. Suppose I observe a raven. It's black. I see a second, also black. And so on for a few dozen more. I reason, or rather I argue with myself, "Since all the many ravens I've seen have been black, the next raven I see will be black." There are seeming problems with this self-argument, this induction-argument. It appears to be invalid since, as is probably obvious, it might be that a non-black raven, perhaps even an albino raven, exists somewhere. And if that's true, then the next raven I see might not be black. Also, the argument is incomplete—as written, though not as thought. As thought, it contains the implicit premise "All ravens are the same color." That makes the entire argument: R = "All ravens are the same color and every raven I have seen was black; therefore the next raven I see will be black." That argument is valid.

Therefore, it is a local truth that "The next raven I see will be black" given those premises. We are back to the same kind of situation as when we discussed Gettier problems. What is our goal here? Is it to assess the truth or falsity of the premises? Or to make predictions? Given the premises are true, then it necessarily follows I will make flawless predictions.

Now "every raven I have seen is black" is true (I promise), so the only question is "All ravens are the same color." Where did that arise? That was an induction-intuition, arising from the judgment that having black feathers is the *essence* of being a raven, or at least part of the essence. If this judgement is true, if having black feathers is essential to being a raven, then this premise is also true and the conclusion to R follows. The crux is thus the step, i.e. the induction, from the observations to an understanding of what it is to be a raven. But there have been observed white ravens, and it is said (by biologists) that these suffer from a genetic *defect*. A defect is thus a departure from the "norm", from what is expected, and what is expected is the form given by the essence. With this in mind we can fix the argument. R' = "All the ravens I've seen have been black and it is the essence of ravens to be black; therefore the next raven I see which is properly manifesting its essence will be black." This is a valid argument, and sound if indeed, as induction tells us, having black feathers is part of the essence of being a raven.

Some people have mistakenly identified features thought to be essential but which were instead accidents. It is not essential that all swans have white feathers; some have black. But because mistakes are made in the induction of essences does not prove that inductions are of no use. Many people make mistakes in math— probably more than who make mistakes in inductions of essences—yet we do not say math is a "problem", where that word is used in its philosophical sense as an unresolved or paradoxical question.

Another example. Who is taller, men or women? Obviously, some women are taller than some men, but everybody knows, via induction from finite observations, that it is the nature or essence of men to be taller. Probabilistically, as we shall learn, it means only that if *all* we know the nature of sex differences and that A is a man and B a woman, the chance A is taller than B is greater than 50%. Not everybody knows *why* this sex difference is so, nor need they know why. It is enough for most decisions and questions to know it exists. Knowing more, we will learn, we can tighten that probability, or come to a deeper understanding of biology. But these are niceties.

We know, via induction, that "men are taller than women", and we know what that phrase means. Goldberg [97, 98] points out that the within-group difference of height in men or women is much larger than the between-men-women difference, but nobody is foolish enough to think this means that men and women are equally tall, or that the small between-group difference doesn't lead to large differences both on average and at the extremes. Yet several very good scientists have been caught making the error that because within-group differences are larger than between-group differences, the between-group differences are, somehow, not there, i.e. that somehow induction has misled us. In other words, because men vary in height considerably, and women too, and because the average difference in height between men and women isn't are large as these differences between men or between women, it doesn't follow that induction has lied to us and that men aren't taller than women. Of course, nobody does make this mistake regarding height differences. The error is usually made on more politically or socially charged questions.

Many applications of induction in science are made at the lowest induction-probability level, where our knowledge is the least surest. Experiments or observations are made to provide the grist for the inductive-probability mill. We observe most men are taller than most women, so we move, via induction-probability, to say it is of the essence of men to be taller than women, which has as shown a probabilistic interpretation. Male-female sex differences in height (and other characteristics) in this probabilistic sense is so firm a piece of knowledge that official "randomized controlled trials" (a term which is highly misleading, as we shall see) aren't run to confirm it. But in most instances in science where experiments are run or observations taken we do not claim knowledge of essential differences. The methods used in classical procedures, both frequentist and Bayesian, are sometimes thought to prove these differences, but we later learn this view is false. We must always come to knowledge of essence via induction, and while experiments and observations can provide the grist they cannot turn the crank of the mill. This is done via intellection. Knowledge does not come via the result of some mathematical calculation such as hypothesis testing. In short, we can never do without induction in some form if we want to do science.

3.3 Grue

The so-called problem of grue (to be defined momentarily) was introduced by Nelson Goodman in 1954 [101] as a "riddle" about induction, a riddle which has been widely thought to cast doubt on the validity and rationality of induction. That unnecessary doubt in turn is partly responsible for the reluctance to adopt the view that probability is part of logic: for that view see, e.g., [35, 79, 122, 206]. Several authors have pointed out deficiencies in grue, most notably [102, 208]. Nevertheless, the "problem" still excites some authors, e.g. [194] (and references therein).

Here, adapted from Groarke (p. 65), is the basis of grue, along with another simple demonstration that the "problem" makes no sense (Groarke lists others, as does Stove). Grue is a predicate, like green or blue, but with a built-in *ad hoc* time component. Objects are grue if they are *green* and observed before (say) 21 October 1978 or *fast* and observed after that date. A green grape observed 20 October 1978 and a fast (say, white) car observed 22 October 1978 are grue. But if you saw the green grape after 21 October 1978, or remember seeing that fast car in 1976, then neither are grue. The definition changes with the arbitrary date.

Imagine it's before the Date and you've seen or heard of only green emeralds. Induction (of what type?) says future, or rather all unobserved, emeralds will also be green. But since it's before the Date, these emeralds are also grue, thus induction also says all unobserved emeralds will also be grue. Finally comes a point after the Date, and lo, a green and not a fast emerald appears, thus not a grue emerald. Induction, which told us that emerald should be grue, is broken!

Why is this wrong? The reason we expect (via induction) unobserved emeralds to be green is we expect that whatever is *causing* emeralds to be green will remain the same through time. Whether this is the formal, material, efficient, or final cause depends on the perspective one takes, of course, but unless there is other specific information, we expect constancy of cause. We comprehend the *essence* of what it is to be an emerald is unchanging. And that is what induction is: the understanding of this essence, an awareness of cause. Rather, that is one form of induction, as we now know.

Nobody has ever seen a fast emerald; neither are blithe, winsome, electrifying, salty, nor brutal emeralds observed. Nobody has ever seen a blue one either, yet it is blue that is the traditional alternate predicate stated in the "problem", not fast or blithe, etc. The choice of alternate predicate is arbitrary; there is nothing special about blue. Using an absurd one like fast makes the so-called problem of grue disappear, because we realize that no emerald can suddenly change nature from green to fast. That is, our understanding (via induction) that it is the essence of emeralds to be green, that some thing or things are causing the greenness, is what leads us to reject the idea that this cause can suddenly switch and create blithe or fast emeralds instead of green ones.

Incidentally, there is no causation in the predicate grue, as has often been noted. Which is to say, the riddle does not suppose emeralds are changing their nature (meaning no change in any formal, material, efficient, or final cause takes place), but that induction is supposed to indicate that some change in nature *should* take place on the Date but doesn't. After all, some thing or things must operate to cause the change. Grue, then, is a mix up in understanding causation.

Again, we do not know of any cause (or any type) that will switch emeralds abruptly from green-mode to blue-mode or to fast-mode. It is thus obvious that the predicate blue is what caused (in our minds) the difficulty all along. We observe that colors change in certain objects like flowers or cars. Via induction, we expect that this change is natural or is of the essence of these objects. Why? Because we're aware of the causes of color change which make the object at one time this color and at another time that color. For instance, a leaf changing from green to red on a certain date. This does not shock because we are aware of the cause of this change. Amusingly, if we re-create the grue "problem" for the leaf using green and red, and we get the right date, then grue-type induction *works* for autumn leaves.

There was never anything wrong with induction. Far from causing us to doubt induction, thinking about grue strengthens the confidence we have in it because we realize that grue seemed problematic because it tortured our understanding of what caused emeralds to be green.

Groarke calls belief in Goodman's grue "an adamant will to doubt rather than an evidence-based example of a deep problem with induction" and likens it to the fallacy of the false question (e.g. "Have you stopped using p-values yet?"). Groarke says (p. 65):

> The proposition, "emeralds are grue," [if true] can be unpacked into three separate claims: emeralds are green before time t (proposition$_1$); emeralds are blue after time t (proposition$_2$); and emeralds turn from green to blue at time t (proposition$_3$). Goodman illegitimately translates support for proposition$_1$ into support for proposition$_2$ and proposition$_3$. But the fact that we have evidence in support of proposition does not give us any evidence in support of all three propositions taken together.

What does the arbitrary time have to do with the essential composition of an emerald? Not much; or rather, nothing. Again, the reason we expect (via induction) unobserved emeralds to be green is we expect that whatever is causing emeralds to be green will remain the same. That is, the essence of what it is to be an emerald is unchanging, and that is what induction is: the understanding of this essence, an awareness of cause. Groarke emphasizes that the time we observe something is not a fact about the object, but a fact about us. And what is part of us is not part of the object. Plus, the only evidence anybody has, at this point in time, is that all observed emeralds have been green. We even have a chemical explanation for why this is so, which paradox enthusiasts must ignore. Thus "there is absolutely no evidence that any emeralds are blue [or fast or hyperbolic or tall or etc.] if observed after time t."

Somewhat related to induction and grue is Kripke's quus example, which I'll hold off on until discussing under-determination in Chap. 7.

Chapter 4
What Probability Is

"He's alive Frank, though he's on life support. Doctors say he's got a 50–50 chance of living...though there's only a 10% chance of that."—Captain Ed Hocken in *Naked Gun*.

Probability is, like logic, an argument. Logic is the study of the *relation between propositions*, and so is probability. Like logic, probability is not a real or physical thing: it does not exist, it is not ontological. It cannot be measured with any apparatus, like mass or energy can. Like logic, probability is a measure of certainty, where by custom non-extreme certainty is called probability; extreme probability, certainly false and certainly true, are thus a special case of probability, probabilities of 0 and 1.

Probability is widely misunderstood for two main reasons: the confusion between ontological and epistemological truth, and the conflation of acts or decisions with probability; [188] also insists on this distinction. These errors are discussed in the next chapter, but it's helpful here to have a précis. A thing is ontologically true if it exists, and it is ontologically false if it does not. An epistemological truth is when we know, given certain unambiguously specified evidence, that a proposition is so. Epistemologically true propositions do not have to be ontologically true. We know the proposition "Mike is green" is true given "All dragons are green and Mike is a dragon". This is an epistemological conditional, or local, truth. But we also know the major part of the premise is ontologically false because there are no dragons, green or otherwise. That we know it is ontologically false—that we know there are no dragons—is itself both an ontological and an epistemological truth, conditional on observation. The lack of dragons is ontologically contingent. The subjects of contingent propositions can only be or not be, exist or not, they cannot have ontological probabilities: they can only be ontologically true or false. Yet since we might not know whether a thing be or not, propositions can and do have epistemological probabilities. Probability is the simple extension of logic to situations where the evidence does not guarantee epistemological certainty.

The truth or probability of any proposition is conditional on the given or accepted premises. We know that "All men are mortal" is true because our

© Springer International Publishing Switzerland 2016
W. Briggs, *Uncertainty*, DOI 10.1007/978-3-319-39756-6_4

senses confirm individual instances and because induction-intellection provides the certainty. This is also why we know "For all natural numbers x and y, if $x = y$, then $y = x$" is true. Individual instances are observed, and induction supplies the rest. We know that "Chicken-man wears a costume" is true if we accept "All super heroes wear costumes and Chicken-man is a super hero." But we also know that the same proposition is false if we accept "No super heroes wear costumes and Chicken-man is a super hero." And we know that "The probability Chicken man wears a costume is 32%" is true if we accept "Just 32% of super heroes wear costumes and Chicken-man is a super hero."

Counterfactuals are always ontologically false; i.e. they begin with premises known observationally to be false. Yet counterfactuals can have meaningful (epistemological) probabilities. The conclusion "Germany won the war" is likely true (we may say) given "Hitler decided not to attack the Soviet Union" and other premises (those which follow) which are plausible or deducible given that premise. Counterfactuals are surely meaningful epistemologically but never ontologically. Counterfuturals, a neologism, are ontologically unknown, but also valid probabilistically. A counterfutural, in relation to a proposition of interest, is a premise which assumes what will be ontologically true but which might not eventuate. Given "Assuming mom comes for dinner" then "I will make cookies." In other words, a counterfutural is a standard, run-of-the-mill prediction, which nobody disputes can be handled with probability.

There is no such thing, therefore, as an unconditional truth or probability. Everything follows from this simple fact. The truth or probability of any proposition changes depending on the evidence. We could not say anything else *but* that "The probability Chicken-man wears a costume is 32% *and no other number*" is true, given *only* "Just 32% of super heroes wear costumes and Chicken-man is a super hero" *and* the tacit premises about the meaning of the words, the grammar, and our implicit understanding (formed via induction, too) about how sets of propositions like these are related to one another.

4.1 Probability Is Conditional

As we saw in Chap. 1, all truth is relative in the sense that all propositions which are (epistemologically) true are so because of some reason or reasons; see [143], except that I do not endorse the use of "inductive probability" because, as we have seen, induction is a many-faceted thing. Also see [170]. All things that exist are also ontologically true given some reason (some cause), incidentally. The propositions of which I speak include necessary truths, which are absolutely true propositions given a set of premises which, if not themselves indubitable, are the valid result of reasoning from inductions using sound rules. We distinguished between necessary truths, which are universally or always true, and locally conditional truths, which are propositions accepted as true when reasoned from given but not necessarily true premises. An example of a necessary truth is the proposition "P is P and not not-P" (where P itself is a proposition, and we include, as always, the tacit premises provided by our understanding of grammar and so forth). This is the so-called Law of Identity. A conditional truth is the proposition "George wears a hat" conditional on the premise "All Martians wear hats and George is a Martian." That probabilities are conditional is recognized in many of the authors mentioned, but also, after a fashion, in [1, 99, 124].

This difference between necessary and conditional truths is no small distinction, as we'll soon learn in the area called "subjective" probability. Most workaday or ordinary truths, and all scientific ones, are conditional truths, propositions which are not necessarily true because the premises which support them are not themselves necessary truths. In most of regular life, and in science, we argue with contingent premises, therefore we can do no better than conditional truths. The constant danger is that conditional truths are taken for necessary truths; and when this happens it is often because of scientism. I'll prove this with examples in later chapters.

Since probability is like logic (or *is* logic), before we can understand the probability of any proposition we have to know which premises are given to support that proposition. In other words, there is no such thing as an unconditional probability. Many authors think there might be; for example Hájek [105]. These authors speak of the supposed difference between epistemic, aleatory, factual, stochastic probability, among other terms: factual or physical probabilities are thought to exist. Now *to exist* is an ontological and not an epistemological statement. If factual probabilities existed ontologically they could, at least in principle, be measured, like mass or electric charge can. Dice throws and human births are often given as examples of things that have factual probabilities. The chance of a boy, we hear, is something like 51 %, a number which comes about by measuring actual births and tallying the relative frequency of boys.

But there is no such thing as factual probability. Each birth is *caused* to be a boy or a girl by some thing or things. Among these things are the genetic makeup of the father's gametes, the conditions in the mother's uterus, environmental stressors and so forth. Each baby is produced by slightly different causes which vary for many reasons. The process itself is constrained to follow certain lines. Human mating doesn't produce turtles or kitchen mops: miniature human beings are made. That we don't know precisely what *all* these causes are in *this* birth is one question, and a good one for biologists and families-to-be, but the causes must be there. Our (epistemological) knowledge of the proposition "It will be a boy" can of course be informed by what we know of the causes and of what we have seen in previous births. But there is no ontological probability *driving* any birth this way or that, a probability which somehow "balances the books" across all of humanity so that just 51 % of boys are born. To think there is to commit the gambler's fallacy.

Maybe this is better seen in a simpler example. Suppose we have a device into which marbles are dropped and which must come out slot A or B, and that the hole through which the marble drops is wider for slot A. Given this evidence and our common experience of falling objects, our epistemological probability is that the marble is more likely to show in slot A in one drop or averaged over many. But each marble as it bumps and spins and drops its way through the device is *caused* to take the path it does. If we knew the characteristics of the machine and the initial conditions of the drop and the equations of motion for objects like marbles, we could *deduce* which slot the marble *must* go through (as we can with, say, coin flips). The answer would be both an (eventual) ontological certainty (for the marble *to be* at this or that slot) and an epistemological certainty.

The exception to this appears to be for quantum mechanical events; photons going through this and that *or* this or that slot, the time at which a nucleus disgorges an alpha particle, whether this particle is spin up or down and its entangled pair the opposite, and so forth. Experiments repeated under very similar conditions—one cannot claim absolute similarity except by induction; if one could know *all* conditions, then quantum mechanics would not be mysterious, but do not forget *all* is the most demanding word there is—show stable frequencies of outcomes. But so would repeated experiments with the marble in the machine, as do human births, coin flips, and so forth. The equations which give rise to the quantum mechanical experiments are probabilistic in nature; they speak of probabilities and not certainties. On the other hand, so do equations saying things about coin flips. Yet in everyday, schoolyard coin flips or all quantum mechanical experiments, all we know is that we don't know the cause which gave rise to the observations. We do know, or we should know, that some thing or things must have caused the outcome. With our marble or with coin flips, being so exposed to the world, as it were, we have some ability to know what these causes are. But this isn't so in quantum mechanics, where the causes are hidden from us. But QM *is* different: in QM we know we can't know *why* any QM experiment happened, but that something caused the outcome we do know. That we know, or should know, causes exist and the nature of causes is discussed later in the chapter on Causation (which includes a discussion on Bell's theorem for QM). All that is important here is that our epistemological understanding (or ignorance) of causes are conditional.

Now to say, "The probability of X is p", where X is some proposition, is to speak incorrectly. Colloquially we often do say things like that, but there is always and must be tacit evidence behind the p. A man says, "The Tiger's are probably going to win today." The man who says this assumes his audience shares the same or much of the information about the game he does, and if this audience does not, the man might use the statement to launch into an analysis, which is a listing of his premises (many of which will be vague) and how they are probative to the proposition of interest "The Tigers win today". Or the listener may say nothing but might, for instance, supply the premise "I trust this guy" and therefore agree with p. The trust is the condition or premise which must exist if he agrees with p and does not know the man's other premises.

To speak properly one must say, "Given this evidence, the probability of X is p." This is why there is no such thing as a probability of being struck by lightning or of dying from a heart attack or whatever. Probability is not intrinsic; there must always be conditions. No probability exists ontologically, and therefore no probability can be calculated unless it is conditional on some evidence. Probabilities speak only of the possible truth or falsity of propositions. Thus (as will be proved below) the probability of "Pat the cat shot at least 30 rounds" given "50% of cats shoot at least 30 rounds and Pat is a cat" is 50 %, but the same proposition has probability 60 % given "60% of cats, etc."

Our job in probability, so to speak, is to figure which side of the equation we're on. All probabilities fit the schema Pr(P|Q), where P is the proposition of interest and Q are the premises, evidence, data, "for granteds", or whatever you wish to call them. The Q must always be there; and so must the P. Now it could be that we could deduce a P from a given Q, like we do in syllogisms, but this needn't be the case. Given Q = "All men are mortal and Socrates is a man" we could deduce P = "Socrates is mortal." But given that same Q we could equally ask the probability of P = "Socrates would have liked Oldsmobiles." The choice of P is ours. This P is, of course, silly, but probability, like logic, does not give any guidance from where or whence propositions arise. This point is basic: reflect on it. Logic and probability speak about the connections between propositions, and not the propositions themselves. This choosing part of probability really is, or partly is, subjective, because the choice of P and Q is sometimes free, and sometimes supplied to us by outside agencies. But once P and Q are fixed—*however* they are fixed— probability is *no longer subjective*, and is itself fixed. This will be proved in the next chapter. For now, we need to recognize there are two actions we can take.

The first is in finding the best or "good" Q for a fixed P. Suppose P = "The Tigers win tonight's game." This is evidently a contingent proposition, in the sense there is (unfortunately) nothing in the universe which would make P necessarily true. After the game is played, we can use the observation Q = "The Tigers won; hooray" and then the probability of P is 1 conditional on this evidence (the event is then also ontologically true). But before the game, what is the best Q? One possible Q is the tautology, "The Tigers will either win or they won't." From this, Pr(P|Q) is the unit interval, i.e. (0, 1). This is because (see below, and as we learned in Chap. 2) tautologies or any necessary truth cannot be used to deduce contingent propositions. Q contains implicitly information on the contingency of P, but that is *all* it contains. Another Q might be, "The Tigers won 4 of their last 5". This gives some idea that P is likely, but we cannot deduce a single, fixed probability from this Q. This non-quantifiability is true of most situations in life. It is only because of scientism that we believe we can impose numbers on everything. Now it may be that this 4-out-of-5 is the only Q we have or were provided. In that case, we're done. The best we can say is the probability of P with this Q is (something like) "pretty good" or "fair." No numerical probability is possible; or, at least, not without tacit premises which insist exactly what "4 out of 5" means regarding future events. Of course, some numerists (as we might call them) jump on examples like this and readily supply premises which become models. More on this later.

It could also be that we have an opportunity, within our means, of searching for additional evidence which is probative of P. This may be the scouting reports on both teams' pitchers, whether the game is home or away, a longer record of wins and losses, injury reports, the weather, batting records, sports writers' opinions, the consequence of the game (is it for the playoffs?), and on and on. There is the sense that as Q increases in richness or content the probability of P should head toward 0 or 1, as the case may be.

It is far from clear that in all cases, particularly those dealing with human behavior, that a realistic Q can be found which puts P arbitrarily close to 0 or 1.

Man has free will[1] and acts unpredictably. But if we had the experimental set up with the marble and wanted to know P = "The marble will arrive in slot A", we have a much better sense, as described above, that we can discover a realistic Q (given precise measurements, say) such that the probability of P is very close to 0 or 1. The *realistic* qualifier is there because we can always and trivially produce a Q which puts the probability of any P at 0 or 1; e.g. Q = "P is impossible" or Q = "P must happen". Realistic evidence is that which itself is ontologically true or possible or necessarily true epistemologically (conditioned on still yet other acknowledged premises). Think of realistic evidence as that evidence which you must justify to outsiders dubious about P. For the marble, this includes the initial conditions, equations of motion, the gravitational field, friction, and so on. Given enough time and effort, we have the idea a pretty accurate Q can be discovered. The search for "Qs" for fixed "Ps" is what science is all about. And, as said, there are some P, such as in QM, for which no Q which leads to certainty will ever be found.

In our daily experience, say when we're flipping coins to solve minor disputes, we don't have the information necessary to find a Q beyond "There is a two sided coin marked H and T, one side of which must show when flipped." This is why we say the probability of an H is 1/2. It is with respect to *that and only that* evidence, but not to the evidence of the physicist who has taken pains to measure the coin flip "experiment" precisely. About the *real coin* in front of us we know nothing *except* that it shares the two qualities mentioned in the premises. It shares in the essence of other backyard coin flips. These premises become a model, a subject which is discussed more later.

The second action is to play detective, to discover the P which best accords with the given, fixed Q. Let Q = "A murder was reported at the mansion, in which resided the Duke and his lady, four aristocratic guests of varying and known histories, and twelve servants. The de-monocled body of Lord Wistful was found strangled in the library. The layout of the mansion and grounds are this and such. The train schedule and distance of the station from the mansion are in this table, etc." In other words, all the standard clues British television detectives are taught to collect at crime scenes. Now suppose you entertained P = "The President of the United States did the deed (though he probably had a good reason to do so, for he is a member of the party for which I habitually vote)." The president was not a guest of the Duke, and was not known to be within thousands of miles of the mansion. Still, P is contingent. Given this Q, which does not exclude any human being by name, it is possible that P is true, though it is very unlikely (recall we are *accepting* Q, which would include premises about how anybody could have done the deed). As above, given Q and including the tacit premise about contingency, the probability of "The President did it" is again the open unit interval.

Clearly, there are better suspects; e.g. P = "The butler did it." Note carefully that Q is *not* "There are 18 suspects one of which is the murderer and the Butler is one of the suspects." *Just* on that, and *only* that, evidence, the probability of P = "The

---

[1] If you say you don't, you contradict yourself.

Butler did it" is 1/18. Even this Q does not limit us, if examined closely; there is a tacit premise that *some* human being in the world did the deed, but that is it. Q is hardly definite, but it is definite enough for the police to begin their interviews. The police would not restrict themselves to the eighteen inhabitants; somebody unknown to the household could have committed the crime. With this Q, no P allows of a numerical probability, but given the nature of Q, some P will be more probable than others. "The butler did it" is an obvious one, and so is "Mrs Duke did it." But those propositions flow from tacit premises in Q given evidence of past murders and so forth.

Of course, the additions and culling of both sides of the probability equation are more natural than my stilted prose has it. As the murder investigation continues, Q is amended and the list of P's tightens. This happens in scientific investigations, too.

So all probability is conditional. My experience is that after introducing this notion, those trained in classical probability and statistics are suspicious and reject the truth that all probability is conditional. The following exercise is usually convincing: try and write down an unconditional probability. Notation is a problem. Many write things like "$\Pr(P)$", which appears unconditional, but that's only because the Q is removed off site. It's always there, however. The danger is reifying the notation so that the conditions Q are thought not to be there *because* they're not written. Once you try this, you'll quickly become convinced. Hint: be careful not to forget the tacit premises of grammar and logic, a negligence which often convinces students they have finally "discovered" an unconditional probability.

Franklin [79, 82] wonders whether there are probabilistic relations between necessary truths. For instance, all the relevant facts, i.e. premises, might be known which support a mathematical theorem, i.e. a proposition of interest such as the Riemann Hypothesis, which, at this writing, remains unproved. It is not known, given the premises mathematicians have today, whether the theorem is proved or not, yet we would think that because all the necessary premises are there, and they really do logically imply truth of the theorem, but we cannot see how, that logical probability might be incomplete. In other words, why isn't the Riemann Hypothesis given all the premises we have, premises which would presumably be the same once somebody comes through with the formal proof? If logical probability were complete, we should see the probability of the theorem was 1. But there is a flaw in this thinking. The easy answer is that any set of premises are not only the premises themselves, but include any number of tacit premises, as we have been assuming all along. For instance, consider you come upon a pile of unconnected electronic parts, resistors, capacitors, wire, solder, and so forth, and the proposition "This is a radio" comes to mind. All the parts for the radio are there, so the answer is that the proposition is true, sort of. The parts are there but the tacit premises of how they fit together are missing. Same thing with the mathematical theorem. We might have all the relevant premises, but how they fit together and in what order is not known. After the theorem is officially proved, it later becomes clear that the facts were necessary and sufficient and these are wondered over, yet how they fit together isn't considered as important. But the form, or the how of all fits together, is also a premise. That is currently missing for the Riemann Hypothesis. This state of mind is natural because talking of the fitting together adds a mental burden when all one wants to convey is the importance of the theorem. Nevertheless, when discussing probability, like in discussing logic, tacit premises should never be forgotten.

4.2 Relevance

We need the idea of *relevant* and *irrelevant* evidence. Irrelevance occurs when the probability of some proposition, given whatever evidence we already have, is not changed in the face of different and new evidence. The probability of P would obviously not change it we merely repeated evidence in Q which was deducible from extant evidence. That is, the probability of P given Q is identical to the probability of P given "Q & Q"; and if Q logically implies W, then the probability of P given "Q & W" must equal the probability of P given Q—but it is not necessarily the case that $Pr(P|Q) = Pr(P|W)$.

Causality and its lack also plays a role. For instance, the proposition "This cargo ship is over 100,000 tons" has high probability given the evidence, "Most cargo ships are over 100,000 tons, and this ship is a cargo ship." If I add the true premise "Einstein enjoyed tobacco" the probability doesn't change: the evidence is irrelevant *because* we cannot identify any causal connection, no matter how complex, between the propositions. This supposition falls short of proof, naturally, but that's because we are in the realm of the contingent. I cannot prove to you that Einstein's hobby is unrelated to cruise ship weights; nevertheless, induction demands it. There is no identifiable logical connection between cargo ships and Einstein's smoking. Induction lurks here as everywhere. The reason we know Einstein's smoking has no bearing on the question is because of induction.

The opposite of irrelevant is *relevant*. If I add the premise "This cargo ship is smaller than most", the probability of our proposition changes: by how much is not known, of course, but it is obvious, given our (tacit) understanding of the English language, that this new evidence is probative, that it is relevant, and it is relevant because it is determinative.

It makes a difference when evidence is introduced. If P = "John was the killer" and we began with Q_1 = "The murderer was a dentist and John is a dentist" the additional minor premise Q_2 = "John graduated from Honest Ben's Dental College" is irrelevant because we already know John is dentist. But if all we had was Q_3 = "The murder was a dentist", adding Q_2 (and leaving out Q_1)*is* relevant. This point is taken up again in models (especially time series). Relevance is itself conditional on the accepted premises.

Evidential importance or *weight* to a fixed proposition is measured, when it can be measured, but how much the probability of a proposition changes on the addition of new evidence. Importance is thus also relative to the information already present in the premises. If we start with Q = "At least 6 of the 11 balls are orange, and this is a ball", the probability of P = "This ball is orange" can be calculated (see below). If we add to Q the evidence "There are 6 orange balls", the probability of

P conditional on the augmented Q is 6/11. How much has the probability "moved"? Clearly by some appreciable and happy amount. We have reduced the choices from six to one. Weight will not always be quantitative, just like probability is not always quantitative. For instance, we could start with Q = "Some of the 11 balls are orange, and this is a ball". As before, we revisit this technique when we discuss modelling.

How relevancy, irrelevancy, importance, and weight are measured, using for instance the techniques of entropy and the like are, of course, of tremendous practical interest. But those techniques, while interesting, are incidental to the philosophical points made here. Jaynes [122] is an ideal reference work.

4.3 The Proportional Syllogism

Given Q = "An Metalunan interocitor must be in one of n states, $s_1, s_2, \ldots, s_n$ and S is a interocitor" then the probability Q = "S is in state s_j" equals $1/n$ (a tacit premise here and throughout is that n is finite; happily, in real life n always is). This value arises from the *proportional* or *statistical syllogism*. Carl Hempel originated the idea in statistics, and it is stressed in philosophy in the works of Stove [204, 205], Williams [224], Groarke [102], and Franklin [79]. The proportional syllogism arises in the natural way from recognizing the equality of probabilities in equations like this, called the symmetry of logical constants:

$$\text{Pr}(\text{S is in state } s_j|Q) = \text{Pr}(\text{S is in state } s_k|Q); j, k \in 1 \ldots n. \qquad (4.1)$$

There is no proof of (4.1): its truth is intuitive, i.e. provided by induction. Now "intuitive" does not mean every person can be made to understand it (or any proposition!), only that some can. Notice carefully that there is no evidence whatsoever about the interocitor *except* that it can take certain states; especially lacking is any evidence about the *symmetry* of its workings. There may or may not be any; we have no clue. All we have is that the interocitor must take a state with one of n labels (Q insists on this). We have *no* idea how *any* of these states arise. We cannot argue from physical symmetry or indeed use any knowledge from physics or engineering; though some try, e.g. [13]. Except for the assurance each of the n states are possibilities, the workings of the machine are a complete mystery to us and, in fact, can be no better than imaginary because, I shouldn't have to add, there are in reality no Metalunans thus there are no Metalunan interocitors. This is of zero importance, however, because logic is not concerned with reality *but with the relations between propositions*. Recall the French-speaking cat example. Given we know the machine has to be in one of n states, and that is all we know, then the probability it is any one of them just is equal.

Interestingly, Persi Diaconis and ET Jaynes attempted proofs of the statistical syllogism which, they thought, avoided the necessity of the symmetry of logical constants, but it will be shown below that the proofs are circular and rely on the symmetry of logical constants after all. David Stove has a proof which also works but which has a quirk, and it will also be shown.

The proportional syllogism is a deduced principle of probability from the equi-probability of logical constants, as in (4.1). Whether in specific instances the results which flow from it match the results given by some other calculation device or principle, such as maximum entropy, is a nice coincidence, but does not obviate it. These principles, whatever nice or desirable properties they have, are not strictly part of probability. Take for example the premise Q = "90% of Martians wear hats" together with "George is a Martian" then the probability P = "George wears a hat" is 90%, which is also so because of the proportional syllogism; but it is constructed differently. To see that, let GM = "George is a Martian", GH = "George wears a hat", and S = "Sally", etc.; then

$$\Pr(GH|Q \text{ and } GM) = \Pr(SH|Q \text{ and } SM). \tag{4.2}$$

Here we have no idea how many Martians there are, except that there are enough to form an even 90%. Unlike in (4.1), the evidence also changes. Note that $\Pr(GH|Q \text{ and } GM) \neq \Pr(SH|Q \text{ and } GM)$ because Sally could be a human or Metalunan. In general, $\Pr(XH|Q \text{ and } XM) = \Pr(YH|Q \text{ and } YM)$ where X and Y are names. The "symmetry", to use that word loosely, comes in considering that any names (labels) X and Y can be used: there is no information to prefer any name over another, thus the probabilities are equal.

The probability "George wears a hat" given "50% of Martians wear hats and GM" is less than the probability "Sally wears a hat" given "90% of Martians wear hats and SM", a fact which also follows from the proportional syllogism. Just as "George wears a hat" has higher probability given the premise "Most Martians wear hats and GM" compared to the premise "Few Martians wear hats and GM." This example requires the tacit premise that *most* is more than *few*. Numbers aren't needed. The probability "George wears a hat" given "X% etc." is equal to "Sally wears a hat" given "X% etc.", with only the tacit premise, given our understanding of percentages, that X is somewhere in 0–100.

It is a simple principle of logic that if the argument from Q to P is valid, then the argument from "Q and T" to P is also valid, where T is any necessary truth. Intuitively, adding something which cannot possibly be false to Q adds "nothing"; we might say that it doesn't change how P flows from Q; it is like multiplying a simple algebraic equation by 1. Given "All men are mortal and Socrates is a man" then "Socrates is mortal" is true; and nothing changes if we append to the premise a necessary truth such as "A is A".

Common necessary truths are logical tautologies. Examples, "Either Mars now is 12 parsecs from Earth or it isn't," "If it is sunny then I will go swimming which implies I will not go swimming if it is not sunny" (we take the entire proposition here), "Either unicorns like chocolate or they don't" (this is true whether unicorns exist or not), and the standby "P or not-P" (which we know is necessarily true based on tacit premises of logic). That last proposition means if Q to P is valid "Q and 'P or not-P'" to P is also valid. Adding the necessary truth "P or not-P" did not change the argument in any way.

The same conditions hold in probability. If on the argument from Q to P we deduce the probability of P given Q to be some value, then adding a tautology or other necessary truth to Q does not and cannot change this value. Thus

$$\Pr(P|Q) = \Pr(P|Q \text{ and } T). \tag{4.3}$$

And this holds even if $\Pr(P|Q)$ doesn't have a numerical value or if it is an interval; it also holds if T = "P or not-P". A concrete example. As above, given *only* Q = "A Metalunan interocitor must be in one of n states, $s_1, s_2, \ldots, s_n$ and S is a interocitor" then the probability P = "S is in state s_j" equals $1/n$. Take the tautology T = "S is in state s_j or it isn't." T is necessarily true whether Metalunan interocitors exist or not, and true regardless which state any of them might happen to be in. Adding T to our list of premises thus does nothing to the conclusion that P takes probability $1/n$. T, because it is a tautology, does not suddenly change our notion of P, nor does it add any information about it.

Consider the probability of P = "S is in state s_j" given only T = "S is in state s_j or it isn't." Since again T is a tautology it gives no information about P. In particular it is false, though it is often asserted, that $\Pr(P|T) = 1/2$. This is more easily believed when any necessary truth can take the place of T. Let T' = "There are 82 millions ducks in the world or there aren't." Then obviously $\Pr(P|T') \neq 1/2$, but even stronger

$$\Pr(P|T) = \Pr(P|T'); \tag{4.4}$$

and this is so even though neither side produces a unique, single number. The closest $\Pr(P|T)$ comes to a number is the interval $(0, 1)$; note that this does not include the extremes, which is just another way for saying P is contingent. In this sense there is some information in T and T', or rather the words and grammar of T, T', or P, which is usually that P is contingent, and which is enough to exclude complete certainty, but that's all; information in tautologies is thus of (literally) infinitesimal value (when it has any value).

Another version of the tautology is T'' = "S might be in state s_j." Implicit in this is that S might not be in state s_j, thus T = T''. Still another version is T^3 ="S might be in state s_j or s_k" with the same tacit inclusion (though it can be argued this version says only these two states are possible; words matter!). Another: T^4 = "S might be in state s_1 or s_2 or, ..., or s_n", again with the tacit admission they might not. Note carefully that T^4 is *not* equivalent to Q = "A Metalunan interocitor must be in one of n states, $s_1, s_2, \ldots, s_n$ and S is a interocitor" because T^4 *never* asserts that any interocitor must be in *any* state, only that it might be in one or another state, or none at all.

It turns out that Metalunan interocitors are actually that race's version of dice, that $n = 6$, and that each "state" is a side upper most upon tossing. Thus Q really equals "A Metalunan die when tossed must show one of 6 sides, labeled 1 through 6, and S is the side that shows on a toss" then the probability P = "S is 3" equals $1/6$. This also follows from the proportional syllogism. Adding tautologies, I hope is now plain, changes nothing. As above,

$$\Pr(\text{S is } j|\text{Q}) = \Pr(\text{S is } k|\text{Q}); j, k \in 1 \ldots 6. \tag{4.5}$$

And this is so even though we know where are no such things as Metalunan dice. Note that Q does not contain, nor should it, any evidence about "fairness" or "symmetry" or "equally weighted sides" or any other similar words, all of which attempt to say something about the actual objects themselves, whereas probability is only concerned with the connections between propositions and not objects *per se*. How to attach probability arguments to objects is what modelling is, which I discuss later.

Any premises about "fairness" are superfluous to probability, which is to say, to the epistemology of the situation, though they might be important to the ontology. Saying that a Metalunan die is "fair", if it means anything to the epistemology, is no more than a restatement that each side is "equally likely", a conclusion we had already reached with the proportional syllogism. That is, given the premise that a device is "fair" then the probability of equal outcomes is uniform; a circular definition. It is like saying, "Given the probability of X is p, the probability of X is p", which is tautological.

But to the ontology, to call a Metalunan—or any Earthly—die "fair", what else can it mean but to claim that each side is *perfectly symmetric*, even down to the quantum level (or whatever, if anything, is below that)? To call an object *fair*, *symmetric*, *balanced*, *equally weighted* or whatever is to say that *no* inspection would reveal *any* conceivable asymmetry. What a remarkable claim! This pristine state of proportionality, which I suppose might exist in some fanciful physics experiment of the future, is impossible in practice to verify. How do you know, except by great expense and effort, whether any device is symmetric across *all* its constituents? How can you ensure any die or coin toss is "symmetric" or "fair"? How can you even define what that means? How *can* a toss be "fair" except that it is designed to produce equal numbers of heads and tails, or equal numbers of sides, etc.? The answer is obvious; and contra, e.g. [209].

Now it is a *separate* question whether the manner in which any particular, necessarily physically real, device produces more or less uniform outcomes. There can be no *real* tosses of a Metalunan interocitor, but that does not stop us from learning its probabilities. But for a real device, we have to do a lot more thinking. To say a device is "fair" says *nothing* about the mechanism of how that device will register a state. In a real die toss, even if we claim the die itself is "fair", i.e. perfectly symmetric, we have said nothing about how it will be tossed. These are ontological matters; see [50]. There will be a gravitational field, perhaps varying. There will be air at a certain density, temperature, and moisture content through which the die flies. The die will leave some person's hand, perhaps coated with traces of sweat and skin, with a certain spin and momentum; it will have begun its position in the hand in a certain orientation. It will hit the floor or table or whatever at a certain angle, and the floor itself will be more or less elastic and which will give some level of frictional resistance. And this does not exhaust the characteristics of the physical environment of the real toss. Indeed, the number of things which *might* influence the outcome are very large (but not infinite). Experience tells us that most of these things will have scant or negligible effect, but perhaps, for this toss, something happens which gives more weight to a previously unconsidered dimension. Who knows? Let's have no more talk about tossing "fair" dice.

We can extend *fair* to include not only symmetries of the device but also to the environment where the device will be "activated", but as you can now see, this is to say a lot. To the epistemology, nothing changes: we still have a circular definition of the probability. But to the ontology, it is everything. Perhaps, as in highly controlled experiments, we will have a lot of evidence about the physical set up. But often we do not, especially when investigating the behavior of people, which do not act as predictably as dice. It is boastful to say even of a simple coin or die toss that the environment is "fair." But we do not have anything like that level of omniscience when it comes to people. Of course, experience over a great many actual dice tosses show us which environments produce uniform outcomes. Casinos rely on this! That experience feeds into our premises and is then used to deduce probabilities.

So what do we say about the chance *this* real object comes up this or that number? Well, that is the subject of modeling, which we will do later. A brief summary: we begin with whatever clear evidence (premises) we have, judging that some characteristics are important and others ignorable, and then move forward to either make predictions or to experimentation, and after experimentation we produce more predictions.

4.4 Details

The statistical syllogism cannot be escaped, and neither can the symmetry of individual constants from which the syllogism is derived. Yet some authors have attempted escapes. The most noteworthy are Jaynes, Diaconis, and Stove. All were interested in assigning equi-probability to events like die tosses. But since assigning equi-probability, or uniformity, has historically been seen as dogmatic, each author tried to derive the assignment of equi-probability from what they saw as different, less dogmatic premises. These attempts are ultimately failures, as I demonstrate below. Stove's come closest, and indeed has the answer hidden in his effort. This section is necessarily mathematical and could be skipped for those already convinced of the statistical syllogism's utility; though all should at least skim Stove's effort. Our first notion of "parameters" arises in these proofs, too.

The following arguments start with the definite knowledge E that M is contingent and can be decomposed into a finite number of possibilities (like sides in coin flips or states of interocitors, or whatever) $M_1, M_2, \ldots, M_n, n < \infty$.

Jaynes [122] gives a permutation argument in an attempt to deduce the statistical syllogism (he does not call it that), but which relies on an unacknowledged assumption. Introduce evidence E which states that either M_1 or M_2 or etc. M_n can be true, but that only one of them can be true. In the case where M is a coin flip, the result can be either $M_1 =$"head" or $M_2 =$"tail". Thus, $\Pr(M_1 \vee M_2 \vee \cdots \vee M_n | E) = \sum_{i=1}^{n} \Pr(M_i | E) = 1$. At this point, there is no assertion that each of these probabilities is equal, only that the sum is 1. We want to assign the probabilities $\Pr(M_i|E)$ for $i = 1 \dots n$. The set of possibilities is $M = \{M_1, M_2, M_3, \dots M_n\}$. Let π be a permutation on the set $\{1, 2\}$. Let $M' = \{M_{\pi(1)}, M_{\pi(2)}, M_3, \dots M_n\}$. That is, the set M and M' are the same except the first two indexes have been swapped in M'. The evidence E is fixed. Therefore, it must be that $\Pr(M_1|E)_M = \Pr(M_{\pi(2)}|E)_{M'}$ and $\Pr(M_2|E)_M = \Pr(M_{\pi(1)}|E)_{M'}$. Jaynes then makes a crucial step, which is to add to E evidence which states that the total evidence is "indifferent" to M_1 and M_2, i.e.

> if it [the evidence] says something about one, it says the same thing about the other, and so it contains nothing that would give [us] *any reason* to prefer one over the other (p. 39, emphasis mine).

Accepting this for the moment, E then says that our state of knowledge about M or M' is equivalent, including the order of the indexes. Thus, (note the change in indexes) $\Pr(M_1|E)_M = \Pr(M_{\pi(1)}|E)_{M'}$, $\Pr(M_2|E)_M = \Pr(M_{\pi(2)}|E)_{M'}$ and $\Pr(M_j|E)_M = \Pr(M_j|E)_{M'}, j = 3, \dots, n$. Which implies $\Pr(M_1|E)_M = \Pr(M_2|E)_M$: that is to say, equi-probable or uniform prior assignment.

We seem to have proven equi-probability. And this argument is fine if what Jaynes says in the quotation holds. But we can see in it the presence of two tell-tale phrases, "indifferent" and "no reason", which are used, and are needed, to justify the final step. This is just begging the question all over again, for how else could the evidence E be "indifferent"? It cannot mean non-probative or irrelevant. That is, Jaynes has assumed uniform probability (and thus, the statistical syllogism) as part of the evidence E, which is what he set out to prove.

De Finetti has a famous "exchangeability" theorem which states that if an "infinite series" of "variables" exists and the order in which the variables arise is not probative, then a "prior" probability of the states exists. The form of the prior is not given by the theorem; that is, *how* the probabilities are assigned is not stated by the theorem; we know only that it exists. Diaconis [55] investigated finite exchangeability in an attempt to see how assignment might arise (see also [131]).

This argument is more mathematically complicated. De Finetti's theorem, which can be found in many places, e.g. [21], states that in an infinite sequence of exchangeable 0–1 variables there is hidden, if you like, a formal (induced) representation as a probability model with a *unique* measure of the probability model's parameters. The key, of course, is that the sequence *must* be infinite. Diaconis, after showing that some finite exchangeable sequences fail to be represented as probability models with unique measures, goes on to offer a proof for certain other finite exchangeable sequences that do. The word "hidden" was apropos, for in exchangeability arises the concept of parameters (in parameterized probability models), a concept which relies on the existence of infinite sequences. I investigate this important topic in Chap. 8.

Here, I follow [55] as closely as possible, almost copying the theorem as it stands but using my notation; interested readers should consult the original if they desire the details, particularly since the original uses graphical notions which I ignore. Let $\mathscr{P}_n$ represent all probabilities on $M = \prod_{i=1}^{n} M_i$ where $M_i = \{0, 1\}, \forall i$, where M is a finite ($n < \infty$) sequence of 0–1 variables. $\mathscr{P}_n$ may be thought of as the probability models on M: it may be written in coordinate form by $p = (p_0, p_1, \ldots, p_{2^n-1})$ where p_j represents the outcome j where $0 \le j < 2^n$ is the binary expansion of j written with n binary digits. Diaconis gives the example if $n = 3, j = 1$ refers to the point 001. Let $M(m, n)$ be the set of j with exactly m ones. The number of elements in $M(m, n)$ is $\binom{n}{m}$: this much is true—the number of elements in $M(m, n)$ is $\binom{n}{m}$—regardless of what the actual probabilities of any outcomes are.

Now, let $\mathscr{E}_n$ be the exchangeable measures in $\mathscr{P}_n$: $\mathscr{E}_n$ will take the place of the measure on $\mathscr{P}_n$'s "parameters". The theorem is stated thus: $\mathscr{E}_n$ has $n + 1$ points $e_0, e_1, \ldots, e_n$, where e_m is the measure putting mass $1/\binom{n}{m}$ at each of the coordinates $j \in M(m, n)$ and mass 0 at the other coordinates. (Uniqueness of each point in $\mathscr{E}_n$ is also covered, but not of interest here.) How is this theorem proved?

e_n represents the measure of drawing n balls without replacement from an urn with n balls, m of which are marked 1, and $n - m$ marked 0, so each e_n is exchangeable. If e_n can be written as a proper mixture of other exchangeable points, it has the form $e_n = pg_1 + (1 - p)g_0$, where $0 < p < 1$: also, g_1, g_0 must assign 0 probability to the outcomes which e_n assigns 0 probability. But because of exchangeability of the coordinates $j \in M(m, n)$ g_1 and g_0 must be equal. And because the probability for any $j \in M(m, n)$ must sum to 1—and here is the big assumption used in the proof—*the mass of each coordinate is* $1/\binom{n}{m}$.

Clearly, the intuition that gave rise to these particular masses asserted in the proof came from the *fact* that the number of elements in $M(m, n)$ is $\binom{n}{m}$. However, other masses work too, as long as they sum to one and assign a probability of 0 to coordinates not in $M(m, n)$. For example, for $j \in M(m, n)$ assign $1/2m$ for the first m coordinates and $1/(2(\binom{n}{m} - m))$ to the remaining $\binom{n}{m} - m$ coordinates.

The reason that the $1/\binom{n}{m}$ mass was chosen is understandable, but there was no explicit reason for it (other than having the probabilities sum to 1) and the desire for symmetry and the equi-probable assignment. So again, the statistical syllogism/equi-probability is tacitly assumed.

Now Stove's attempt, in my notation and somewhat shortened. The statistical syllogism is deduced from the symmetry of logical constants in this example. Given H = "Just two of Abe, Bob, and Charles are black", the probability of B = "Abe is black", relying on the statistical syllogism, is 2/3. Let T be any tautology, a necessary truth. Then $\Pr(HB|T) = \Pr(H|T)\Pr(B|TH)$. Rearranging, and because logically TH is equivalent to H, we have $\Pr(B|H) = \Pr(HB|T)/\Pr(H|T)$.

H is logically equivalent to

$$B_1 B_2 B_3^c \vee B_1 B_2^c B_3 \vee B_1^c B_2 B_3,$$

where B_1 = "Abe is black", B_3^c = "Charles is not black", and so forth. And that means

$$\Pr(B_1|H) = \frac{\Pr((B_1 B_2 B_3^c \vee B_1 B_2^c B_3 \vee B_1^c B_2 B_3)B_1|T)}{\Pr(B_1 B_2 B_3^c \vee B_1 B_2^c B_3 \vee B_1^c B_2 B_3|T)}.$$

Distributing B_1 in the numerator gives

$$\Pr(B_1|H) = \frac{\Pr(B_1B_2B_3^c|T) + \Pr(B_1B_2^cB_3|T)}{\Pr(B_1B_2B_3^c|T) + \Pr(B_1B_2^cB_3|T) + \Pr(B_1^cB_2B_3|T)}.$$

because $B_1^cB_2B_3B_1i$ is impossible. Here is Stove's big move. He states

$$\Pr(B_1B_2B_3^c|T) = \Pr(B_1B_2^cB_3|T) = \Pr(B_1^cB_2B_3|T); \qquad (4.6)$$

but *also*

$$0 < \Pr(B_1B_2B_3^c|T) < 1. \qquad (4.7)$$

Thus because of the symmetry of individual constants, the statistical syllogism is deduced. The 2/3 probability follows from the labels, here the names, being "exchangeable" with respect to T.

4.5   Assigning Probability: Seeming Paradoxes and Doomsday Arguments

Assigning, or rather deducing, probabilities isn't always easy; indeed it could be formidably difficult, or even impossible. It's sometimes unclear whether, in a problem, the focus is on the P or the Q in $\Pr(P|Q)$. Because of that, disputes about the nature of probability arise. This section shows how tricky assignment can be.

The Monty Hall problem is infamous. Here are the premises: A contestant on a game show is offered to choose one of three doors, A, B, or C. Behind one and only one is a prize, and behind the others there is nothing. The contestant chooses, say, B. Monty then opens, say, C, behind which is nothing. The contestant is then offered to stay with his original choice or to switch to A. What decision maximizes the chance he wins something?

This problem became infamous when Marilyn vos Savant wrote a column in which she revealed the contestant ought to switch, which gives a 2/3 chance of winning the prize. Many—all too many—irate readers wrote in, claiming various superior credentials, including being professors of mathematics. These folks said vas Savant was wrong and that the probability of winning switching or staying was "obviously" 1/2. These professors were, it might surprise you to learn, right. But then so was vos Savant. Because all probability is conditional, people were giving the right answer to the wrong question: they did not see that their premises were different than vos Savant's.

Given *only* the premises, Q = "Before us are two doors, A and B, only one of which conceals a prize", then the probability P = "A has the prize" is 1/2; and for door B, too. Switch or stay; it's the same. But those aren't the premises vos Savant used. She began with the original premises but added to them knowledge that if the contestant initially chose the right door, Monty could open either remaining (thus the contestant should stay); but if the contestant had chosen incorrectly, then Monty could only open the one remaining empty door (thus the contestant should switch). Since picking the right door on the initial premises has a 1/3 probability, vos Savant was right: switching makes more sense. This incident, which attracted nation-wide attention, proves that it's often tough to remember what the premises are and when.

There are many other probability "paradoxes" and problems. The Sleepy Beauty, three envelopes, one child born on a Tuesday, and many, many more, all of which share the same nature as the Monty Hall. Which is to say, they all have hidden, or not readily visible, premises, or they have deductions and implications which are missed, or information that is hideously complex to grasp. But all of them prove that probability is conditional; indeed, it is their conditionality which makes them interesting and (for some of us) fun.

Another class of probability problems also exist, such as Buffon's needle and the so-called marginalization paradox (see Jaynes, Chap. 10 for a gruesome vivisection of this concept), all of which involve making finite choices from infinite sets. These problems also show that probability is conditional. But because messing with infinity is like walking through a raging forest fire and hoping for the best, many folks get burned. Or perhaps it is better to say infinity is like a foreign country; rather, many foreign countries, since there are many kinds of infinities. Mistakes are made when the traveler thinks he has the whole place figured out after only a brief visit. Whenever it is claimed that some "paradox" involving infinity has invalidated this or that philosophy of probability, it is safest to put the claim down to enthusiasm and to continue believing in probability. I say more about infinity when discussing measurement and models. But first a simple example I learned from an unpublished work on probability by Purdue's Paul Draper,[2] also see [63]. There is nothing unique about the example, which is familiar especially in criticisms of Bayesian theory on assigning "prior" probabilities, though Draper phrases it nicely.

Imagine a factory that spits out tiles anywhere from 1 to 3 inches in width. The so-called principle of indifference would lead to a uniform probability assignment to the widths between 1 and 3 in. Since the tiles are square, the surface area is anywhere from 1 to 9 in^2. Considering area under the principle of indifference, says Draper, leads to the assignment of an uniform probability to the areas from 1 to 9 in^2. But then there is a contradiction. Given the indifference criterion and the evidence we're provided, there is probability 1/2 for the proposition "The surface area of this tile is between 1 in^2 to 4.5 in^2", which corresponds to widths 1 in and 2.12132 in. But there is also probability 1/2 for the proposition "The width of this tile is between 1 in and 2 in." What has gone wrong? Many blame probability. Instead, our ideas of measurement have gone awry.

Any real tile can only be manufactured in discrete increments. Call those increments δ. These do not have to be equal, and one length may even depend on another; the only restriction is that $\delta > 0$. For ease, I'll assume this is fixed. That means the widths can be 1 in, $1 + \delta$ in, $1 + 2\delta$ in, and so on up to 3 in (with the

---

[2]Draper, D.: Unpublished manuscript (2014).

additional assumption that for some n, $1 + n\delta = 3$, but again, this is only for ease). Since widths are fixed, so are areas, which can now *only* be 1 in^2, $(1 + \delta)^2$ in^2, $(1 + 2\delta)^2$ in^2, and so on up to 3^2 in^2. No surface area between 1 in^2 and $(1 + \delta)^2$ in^2 is possible. Using the statistical syllogism, the probability "The width of this tile is $1 + m\delta$ in, where $0 \geq m \leq n$" is $1/(n + 1)$. Obviously, since the surface areas are one-to-one with the widths, the uniform probability applies to them as well. No contradiction. And no restriction, either, because we can let δ be as small as we like, smaller even than a quark!, as long as it is non-zero. It is *only* at the *limit* where these odd contractions pop up, but that is because, as already said, infinity is a strange place, and because the limit here is not only δ going to 0, but the n going to infinity in some sort of tandem. How you get to infinity matters.

Another example from Draper. There are three balls in a bag, each of which must be either white or black. Given this evidence, what is the probability "All three balls are black"? Drapers says 1/8 because "Consider the following eight statements: all three balls are black, the first two are black and the third white, the first two are white and the third black, etc. One can easily imagine having no more reason to believe any one of those eight statements than any other." But then, says Draper, "there are also four possible ratios of black balls to total balls in the urn (i.e., 1, 2/3, 1/3, and 0). . .[and] the principle of indifference implies that the probability of the urn containing three black balls is 1/4." Contradiction! Yet Draper forgets some of his evidence. One of the ratios is indeed three out of three, and another is two out of three. But there are three ways to get 2/3: B1B2W3, B1W2B3, W1B2B3. Likewise, there are three ways to get 1/3, and just one way to get 0/3, That makes eight total ratios, only one of which contains all black balls; thus, conditional on the full evidence (and notice even Draper started by labeling the balls but then forgot), we're back to 1/8. Many similar paradoxes resolve in precisely the same way.

Senn [195] has a similar paradox, one common in Bayesian statistics and which causes consternation, using an argument from continuity. He first defines an "event" which can take one of two values, e.g. success and failure. He then *defines* the "probability of success" of this event as θ, and says it in turn is equally likely to take any value in $(0, 1)$ (*sans* extremes). Bayesians would call this putting a "flat prior on θ." He says "Suppose we consider now the probability that two independent trials will produce two successes. Given the value of θ this probability is θ^2. Averaged over all possible values of θ" this is 1/3 (the integral of $\theta^2 d\theta$). The difficulty enters in the next step: "A simple argument of symmetry shows that the probability of two failures must likewise be 1/3 from which it follows that the probability of one success and one failure in any order must be 1/3 also and so that the probability of success followed by failure is 1/6 and of failure followed by success is also 1/6."

Ignoring the language about "events" and "independence", this is a seeming paradox. Why? Because of the odd statement "Suppose that we believe every possible value of θ is equally likely. . ." What can that mean? Nothing. There are more than a simple infinity of numbers of possible values of θ when that parameter is continuous, and to "believe" each is equally likely (based on what premises?) is to claim rather shocking, almost omnipotent knowledge. On this view, what is the probability (still with undefined or vague premises) that θ takes any value? Well, zero. How about after we take some data, consisting of some observed string of successes and failures, what then is the probability θ takes any value. Bayes's rule tells us the answer is the same: zero. This θ is a truly strange creation, a *continuous* string of numbers: no, they are not numbers at all, θ is the continuum itself. And it came out of the blue, with premises that just asserted it. To put a probability to every value of the continuum is to claim knowledge of those values, which is never something we can directly have. The math works out because we have indirect knowledge in the sense of we know the continuum exists and that it has certain properties, but we can never have knowledge of more than a (limited) finite set of those values. Anyway, as is now easy to see after Draper's examples, the paradox disappears if θ is a as-large-as-you-want finite set of values. I'll leave this as homework for the reader to prove. There are, it is suspected, many things to be learned from investigating the continuum, as infinities always boggle the mind. This is why it is right to suspect our limited understanding of the infinite and not our understanding itself when confronted with a paradox.

Next consider the famous Doomsday Argument [17], given in various (anthropic) forms, which supposedly predicts the total number of humans who will ever live. It's also called the Carter catastrophe; the same Carter famous for the anthropic principle, e.g. [11, 146]. The literature is voluminous. The DA is said to be obviously wrong by all who encounter it, but it's also said that everybody's explanation why it's wrong is flawed. Below is a full and correct explanation why the DA fails. The short version is that sloppy notation and forgetting all probability is conditional causes the error.

To solve the DA, the only rule we need recall is this: all probability is conditional—and conditional *only* on the information provided. The idea is that you're born, you notice your birth, and you reason that your place in the order of all human births is nothing special. From that, can we conclude how many more of us we expect? This situation is analogous, at first, to balls in a bag. Our evidence is X = "There are N balls labeled 1 through N in a bag, from which only one will be removed." The probability of Y = "The ball will have label j, where j is from 1 to N inclusive" is $1/N$, via the statistical syllogism. We deduce via the language used that N is finite (no bag can hold an infinite amount of any real thing; this is no restriction, let N increase to a googolplex to the power of a googolplex to the power of a googolplex, 84 times, which is a mighty big number, but finite).

Reach into the bag and pull out the ball B. It will have a label; call it B = j. Our evidence is now augmented: we have *in toto* X' = "X and the ball has label j". What can we say about N? Well, given X', the probability N is less than j is 0, and the probability N is at least j is 1, both of which are obvious. But what about these interesting and relevant probabilities (both given X', naturally): "$N = j$", "$N > j$"?

We do not know. Why? Because there is *no* information in X or X' about the possible values of N, except that N must be at least equal to j (given X and not X'), information which is deduced. Now mentally you might add information that is *not provided*, by, say, thinking to yourself, "This j is awfully low and that's such a big bag; therefore, surely N is large." Or "I know this Briggs, who is a trickster. He made the bag big on purpose. N is small." Or anything, endlessly. *None* of these additions are part of the problem (the stated evidence), however, and all such moves are "illegal" in probability.

Now suppose we legally augment our X and, for fun, say that N is some number in the set S. We don't need to know much about S, except that it exists, is finite, and contains only natural numbers. Thus, X now equals "There are N balls labeled 1 through N in a bag, from which only one will be removed; and N is a number in the set S." Given X, the probability "$N = s_i$ (one of the set S)" is $1/|S|$, where $|S|$ stands for the number of elements in S (its cardinality); thus, the probability "$N = s_i$" is $1/|S|$, where I'll assume the s_i are increasing in i. What about the probability that the ball withdrawn has label j? Here it gets tricky, so let's be careful.

The key lies in realizing the bounds of j are between 1 and the largest value of S. First suppose $N = s_1$. We want: $\Pr(B = j|N = s_1, X)$. This is $1/s_1$ for $j = 1$ to s_1, and 0 for all those j up to s_I (the largest value of S). Now $\Pr(B = j|N = s_2, X) = 1/s_2$ for $j = 1$ to s_2, and 0 for all values up to s_I. From this, we notice we have to be careful about specifying j precisely. From total probability we know

$$\Pr(B = j|X) = \Pr(B = j|N = s_1, X) \times \Pr(N = s_1|X) + \cdots$$
$$+ \Pr(B = j|N = s_I, X) \times \Pr(N = s_I|X),$$

and where knowledge of j is relevant to the probability. If $j = 1$, then

$$\Pr(B = 1|X) = [(1/s_1) + \cdots + (1/s_I)] \times (1/|S|),$$

but if j is a number larger than, say, s_1 but smaller than s_2, then (call this j') $\Pr(B = j'|X) = [0 + (1/s_2) + \cdots + (1/s_I)] \times (1/|S|)$ and so forth for other j (don't forget S is *known*).

The ball is withdrawn and $B = j$. Can we now say anything more about N? As before, there is 0 probability N is less than j, and so if j is greater than some s_i, there is 0 probability N equals those s_i. We can do more, using the good reverend's rule, but it's still tricky:

$$\Pr(N = s_i|B = j, X) = \frac{\Pr(B = j|N = s_i, X) \Pr(N = s_i|X)}{\Pr(B = j|X)}.$$

First suppose $j = 1$, then

$$\Pr(N = s_i|B = 1, X) = [(1/s_i) \times (1/|S|)]/([(1/s_1) + \cdots + (1/s_I)] \times (1/|S|))$$
$$= (1/s_i)/[1/s_1 + 1/s_2 + \cdots + 1/s_I].$$

If you stare at that fraction for a moment, and recalling that the s_i are given in increasing number, you realize that values of smaller N are more probable than larger values. As a for-instance, suppose $S = \{20, 21, \cdots, 40\}$, which has cardinality 21. Given X, the probability "$B = 1$" is $(1/20 + 1/21 + \cdots + 1/40) \times (1/21) = 0.02761295$. Thus $\Pr(N = 20|B = 1, X) = 0.04416451$, $\Pr(N = 21|B = 1, X) = 0.04206144$, etc. out to $\Pr(N = 40|B = 1, X) = 0.01472150$. Notice that these probabilities do not change for j between 1 and 20.

In this same example, next let $j = 21$, then

$$\Pr(N = s_i | B = 21, X) = \frac{\Pr(B = 21 | N = s_i, X)\,\Pr(N = s_i | X)}{\Pr(B = 21 | X)}.$$

For "$N = 20$", the first term on the right equals 0, and so $\Pr(N = s_i | B = 21, X) = 0$, as desired. For "$N = 21$", we have

$$\Pr(N = 21 | B = 21, X) = \frac{\Pr(B = 21 | N = 21, X)\,\Pr(N = 21 | X)}{Pr(B = 21 | X)}$$

$$= \frac{(1/21) \times (1/21)}{([0 + 1/21 + 1/22 + \cdots + 1/40] \times (1/21))}$$

$$= \frac{(1/21)}{[0 + 1/21 + 1/22 + \cdots + 1/40]}$$

$$= 0.06994537,$$

and for $\Pr(N = 22 | B = 21, X) = 0.06676604$, out to $\Pr(N = 40 | B = 21, X) = 0.03672132$.

Collecting all these tidbits leads to the conclusion that smaller (but not impossible) values of N are always more likely than larger, regardless of the value of j. Why? That's easy. Before we see B, the possible values of N are s_1, s_2, and so on up to S_l, each equally likely. After we see B, some values of N (from S) might now be impossible, but since j will always be less than any remaining possible larger members of S, smaller values of N are closer to j than larger, thus smaller values are more likely. Simple as that.

What does this have to do with Doomsday? Everything. The crucial step was in conjuring the set S. Where did that come from? I made it up. S was known throughout second part of the calculations and unknown through the first part. When S was unknown, N was unknown, and there was *nothing* we could say about N except that it had to be as large as j. I mean *nothing* in its literal, logical sense. In that case, given *only* that you witness your birth order, your $B = j$ that is, we are blind about the future of humanity.

When S was known, we had a rough idea of what N was, which we tightened slightly by learning where N might not be (by removing the ball; by witnessing your birth number). But for an S with large cardinality, we aren't learning much by viewing B. Equal probability on S is what we started with, and something very like equal probability on S is what we ended with. But this is cheating because I made the S up. We wanted N, of which we are ignorant, and then we pretend we know an S that tells us something but not everything about N! All the other solutions to the Doomsday argument I have seen also make up S, but then they add an extra layer of cheating. We posited a discrete finite S, from which deduced that N might equal any of its members with equal probability (before seeing B). But those who conjure up more creative S often fix the set so that smaller values of S are more likely (hence smaller values of N are more likely, even before we see B). Some form of exponential "distribution" for S is popular, though some thinkers create two-member (and not 20-member) sets like $\{10^{10}, 100^{10}\}$, i.e. a low value for all humans and a large value. Some even use non-probability arguments on N (called "improper priors"), which is triply cheating.

Once S is fixed, however it is fixed, the calculations flow in the same manner as above, but it's easy to see that smaller values of N are always going to be more likely than larger, and that's because the j will always be smaller (or no greater) than the maximum value of S. And given that some let S toodle out to infinity, it's no shock at all to discover that N is not expected to be big. Also understand that N is not "drawn" from S: N is caused to be some number, depending on myriads of reasons. Probability represents uncertainty, not cause (as we shall see).

Thus the Doomsday Argument is really a non-problem which includes its own answer in its formulation, which is cheating. Of course, it makes perfect sense to ask the question of how many of us there will be left, but trying to discover the answer using only your birth order is doomed to failure (beyond proving that N must be at least as large as j). Since all probability is conditional on only the information supplied, many different answers for our future numbers are possible. It's easy to think of probative information: demographics, politics, epidemics, apocalypses (giant rocks from the sky, Christ's return, etc.), and on and on. (Of course, some of these sets of information may lead to the guesses people have made about S.) I do not have a good answer how to use these to put uncertainty on (the real) N.

There is more we can say about the errors made in the DA. One difficulty lies in misunderstanding Bayes's theorem, which some mistakenly write like this:

$$Pr(N = s_i | B = j) = \frac{Pr(B = j | N = s_i)\, Pr(N = s_i)}{Pr(B = j)},$$

where the evidence about N in X is left off (finding the denominator is no problem because $Pr(B = j) = \sum_i Pr(B = j | N = s_i)\, Pr(N = s_i)$). $Pr(N = s_i)$ is thus "naked" and violates the rule that all probability is conditional, yet users of Bayes's theorem are trained to posit "priors" like this, and so posit one they do. It seems, say critics of the theory, that these priors are pulled from thin air. The critics are right. It's completely arbitrary to conjure a $Pr(N = s_i)$, and so the resulting $Pr(N = s_i | B = j)$ cannot be trusted.

Of course, I made up my own "prior", but referenced as being a deduction from X. The probability $Pr(N = s_i | B = j, X)$ is thus true. The attention then focuses on X, where it belongs. Why my X? No reason at all. If we're after the best information about N, that is what should go into X. But it has to be information that is not N itself, like my S was. My S was merely a presumption that I already knew a lot about N; it was N by proxy, but a fuzzy proxy. Cheating, like I said. It's not Bayes's theorem that's the problem. It works just fine when we supplied information in X about S. But it also worked dandy when X was just "There are N balls labeled 1 through N in a bag, from which only one will be removed." (The reader should verify this.)

The DA also illustrates the traps of poor notation. There is, as I have been stressing, no such thing as unconditional probability, and in displaying probability when the conditions are dropped, which often makes manipulating the equations easier, you run the risk of introducing error, which is what happens in the standard DA. A standard application of the DA starts by asking for this: $Pr(N < 20j)$ (the 20 comes from the magic number in statistics). Note the missing conditions. Accepting the bare notation, then $Pr(N < 20j) = Pr(N/20 < j) = Pr(j > N/20) = 1 - Pr(j \leq N/20) = 1 - 0.05 = 0.95$. It is said $Pr(j \leq N/20) = 0.05$ because j is "uniform" or is "uniformly distributed", as if probability has life. The fatal error has been made, because we notice that this result appears to hold regardless what value N or j has. But there just is no such thing as $Pr(N < 20j)$. There can, however, easily be a $Pr(N < 20j|X)$.

Here's what we actually want:

$$Pr(N < 20j|B = j, X) = \frac{Pr(B = j|N < 20j, X) Pr(N < 20 \times j|X)}{Pr(B = j|X)}. \qquad (4.8)$$

Now X can be anything relevant; it as least says there are balls 1 through N, but it must also say *something* about N (directly or implied). Suppose X contains information that N is in the set $\{1, 2, \ldots, 19\}$. Then $Pr(N < 20j|X) = 1$ for any j. Never forget j runs from 1 to N, which is where the DA goes wrong: knowledge of N is relevant to knowledge of j, and vice versa. When that is forgotten, anything can happen. It appears, because of loose notation, many forget that j and N are related.

The result of (4.8) is the right-hand-side is 1/1, and thus $Pr(N < 20j|B = j, X) = 1$, as expected. So here is a proof showing that at least one "prior" on N ruins that 95 % finding. Here's another one. Suppose X says $N = 20$. Then $Pr(N < 20j|X) = 0$ for $j = 1$, and $Pr(N < 20j|X) = 1$ for $j > 1$. This amounts to the same thing, that $Pr(N < 20j|B = j, X) = 0$ when $j = 1$, else it equals 1 for all other j. Next suppose X says N is in set $\{20, 21, \cdots, 40\}$. Starts to get interesting. I leave this one as homework.

4.6 Weight of Probability

Many authors following Keynes write about the "weight" of a probability, e.g. [197]. The weight speaks of the judgment of a probability in relation to its evidence, how important that evidence is with respect to the other acknowledged evidence. According to Keynes [132, p. 78]:

> As the relevant evidence at our disposal increases, the magnitude of the probability of the argument may either decrease or increase, according as the new knowledge strengthens the unfavorable or the favourable evidence; but *something* seems to have increased in either case,—we have a more substantial basis upon which to rest our conclusion. I express this by saying that an accession of new evidence increases the *weight* of an argument. New evidence will sometimes decrease the probability of an argument, but it will always increase its 'weight.'

If we have two sets of premises Q_1 and Q_2 which are unrelated to each other in the sense one is not the other with additions, or that one is deducible from the other, where both are probative to some proposition X, then $Pr(X|Q_1)$ might not equal $Pr(X|Q_2)$. Since the two premises Q_1 and Q_2 are not related, one cannot be said to give more or less weight to X than the other. Adding weight implies adding relevant information. For example we might say the probability judgment $Pr(X|Q_1 \& Q_2)$ has more weight than $Pr(X|Q_1)$ or $Pr(X|Q_2)$ alone. This use of weight is uncontroversial and sensible. But there is another sense where weight can be misinterpreted.

Jim Franklin gives the following example [79]. He says the probability of "this coin comes ups heads" given "this coin appears symmetrical and about 500 of the 1000 throws with it have come up heads" equals 1/2. But he also says the probability "this coin will come up heads" given "this coin appears symmetrical" also equals 1/2. Franklin says, "Though the numerical value of the probability is the same in each case, there is plainly some important difference between the two." And this difference is the idea of "weight." The former evidence seemingly has greater weight than the latter, at least, some say, in the sense that any change to the premises of the latter can change the probability dramatically.

We needn't accept Franklin's probability assessments (though they are harmless enough), because it is not clear in the first case the probability should be a single number, and it is the case, as argued earlier, that symmetry or fairness arguments are circular (though symmetry implies two-sidedness here). Anyway, accept the values for the sake of argument. The probabilities are then correct, and given they are correct, both should be believed. They should be believed in just the same way that "Exactly half of all Martians wear hats and George is a Martian" gives the probability of 1/2 to the proposition "George wears a hat." The "weight" of *belief*, and not of evidence, is and should be equal in all these cases.

Perhaps it is not yet clear, so let's imagine three people, C, D, and E. C is a newcomer to the (now dwarf) planet Pluto and has never seen the game of plutonk, which is explained to him as a device which when activated must take only one of two states, s_1 and s_2. Based on this evidence and using the statistical syllogism, C reasons the probability of s_1 equals that of s_2. D is an experienced player and has been standing for several Plutonian days in front of the same machine. He knows the same rules as C does, plus he has seen the device come up s_1 about 500 times out of the last 1000 activations. He figures the probability of s_1 is also about 1/2. If C or D were to bet on the next activation, they would use that 1/2 probability in figuring the amount of their wager (the exact amount depends on factors not relevant to the probability). Both would be figuring correctly.

Enter E, an inveterate and suspicious gambler. Like C, he has never seen this particular device, but he knows the rules. His worry engine starts: "This device might be crooked." Adding that premise to the list changes nothing because it is a tautology: yes, the device might be crooked, but then again it might not be. Simply claiming it might be is to say nothing. But now E adds, "I don't trust it until I've seen it in action." Adding that premise *does* change the probability; indeed, depending on how we interpret it, the probability s_1 may be non-numerical or perhaps some interval. It surely cannot be calculated simply.

Each player has their own probability of s_1, but each player starts with *different* evidence. Since probability is conditional on the evidence, the probability is liable to change with it. As shown earlier in this chapter, there is *no* unconditional probability, hence no unconditional probability of s_1. The feeling that there should be is what accounts for the different feelings of "weight" given the three different sets of premises. We are after that "unconditional" or "true" probability of s_1—and which, in this case, everybody "knows" is 1/2. Weight of belief is in this sense thus a confusion or a false expectation that a true ontological probability exists for every unique situation.

There is another interpretation of weight of evidence, however. If we start with C's premises, we can make predictions. I'll not do those here; I will when we do modeling. Anyway it's obvious predictions can be made. Before we see the device in action, the predictions will be imprecise, based on C's premises. After we add evidence like D's, then our probability will change and our predictions will become sharper. The increase sharpness is a measure of the weight of the observational evidence. After 1000 observations, it is not likely the next observation, or indeed other evidence, will change the probability much. New observations are thus accorded little weight, where weight is now the difference (of some kind) in probabilities deduced for the two sets of premises. Jaynes provides a wealth of mathematical details about this idea in his "A_p" distribution. This version of weight is like importance described above, and is relative or conditional.

This kind of weight accords with tradition and common sense and is why we say (things like) extraordinary claims require extraordinary evidence. The weight of new evidence for startling claims has to be very large to overthrow, say, centuries of tradition.

4.7 Probability Usually Is Not a Number

If the evidence is "Most Martians wear hats and George is a Martian" then the probability of "George wears a hat" is not a unique number. This is because of the ambiguity of *most*. There is no information in the premise to say precisely what *most* means. Nevertheless, using its plain English definition—I mean the one we carry in our heads and not any official dictionary's version—*most* implies, to my ear, *at least half but not all*. Given the original premise and this tacit one, then the probability of the proposition is greater than 1/2 but less than 1, an interval. But other tacit premises are possible. Of course, we haven't yet proved that probability can (sometimes) be represented numerically, but accepting that it can, we have at least one case where probability is an interval. Note carefully that if we do not have this tacit definition of *most* in hand, then there is *no* probability; and even when we do, the probability is not a unique number.

Switch to "Some Martians wear hats, etc." and the probability becomes fuzzier. *Some* has a colloquial and a logical definition; either can be used, and the result depends on what you bring to the word. In my ear, it means *not none and not all*.

That is *not* the logical definition, which extends to *possibly all*. But using my tacit premise gives a probability greater than 0 and less than 1, i.e. the uncapped unit interval. I don't mean this is a subjective interval: I mean that if *some* takes the definition *not none and not all* then the probability of "George wears a hat" is *necessarily* the uncapped interval between 0 and 1. If you accept the given and the tacit premise then you too must agree with this probability, just as you must agree with the probability of 1 when the premise is "All Martians wear hats, etc." If you find yourself rebelling at this, wait until the discussion of so-called subjective probability. Again, there is no movement *unless* we supply the tacit definition.

Of course, we always do and must bring these tacit premises: we bring to every argument our understanding of all its words and grammar. It is the same understanding you bring to understanding this sentence.

Suppose the premise is "David, who is 5'8", is 205 pounds." Given that, what is the probability "David is fat"? There is no answer given *just* this information because there is ambiguity in *fat*. The word incorporates sex, height, weight, age, culture, any number of personal biases. But there is the sense that if we had the specific definition of *fat*, especially with regard to sex (David is a man's name), height and weight, then the probability "David is fat" is either 0 or 1, depending on whether the premises matched the definition, and accepting the other dimensions (age, culture, etc.) have no bearing. Yet we do not have this definition, so there is no probability. We could supply them, as we did with *most* or *some*, but they are not part of the given evidence. Supplying them changes the problem. And there surely is less agreement as to what constitutes *fat* than *most*.

Ambiguity can appear in both the premises and the proposition of interest. Let "There is a humongous number of balls in the bag"; given that, the probability of "There is a boat-load of balls in the bag" is undefined. What is *humongous* to *boat-load*? Are they equivalent? Is one higher than the other? Who knows? There is no dismissing this example on the grounds of ambiguity because, as we all know, much communication is in this form. People in their daily lives are often satisfied with imprecision; indeed, why should they not be? What advantage in forced quantification outweighs the burdens of investigating it most of the time? In highly technical areas the answer is obvious, but for most of life it surely is not.

We often hear and speak sentences like, "It's probably not going to rain," "I think the Tigers will win," "It isn't too likely she'll come", "It might happen", "It's a possibility", "It's a probability." The latter two are distinct in British English but not so much or at all in American English, so we have to be careful about misinterpretation, as always. The propositions of interest to which these sentences speak are easy to extract: "Rain", "The Tigers win", and so on. The evidence or premises used are hidden, and are more or less known by the listener. Locutions like this acknowledge the non-quantifiable nature of probabilities with respect to certain evidence; see [147]. And unless we have a crucial decision riding on the proposition, like a bet, then we're happy with this level of fuzziness. Although I don't here discuss it, "fuzzy logic", where it is coherent and relevant to questions of uncertainty, is a rediscovery of probability or probability by another name (this was also Martin Gardner's view, [87]).

Some have conducted experiments to see what numerical probability is most associated with words like *likely*, *maybe*, and so forth, e.g. [47]. The results from studies like these are quaint and of some interest to, for instance, linguists, but they provide no philosophical insight. We might learn, of this certain kind of people, that when they hear "likely" they interpret it to mean numbers in the range, say, 0.8 to 0.9. But unless we knew what a specific person was thinking about a probability word, it doesn't help. If we knew the person belonged to the certain group, we might be able to specify an interval.

Legally, guilt in capital crimes is (or was) meant to be "beyond reasonable doubt." The jurors hear the evidence Q, and this is augmented with their own ideas, prejudices, beliefs, experiences, which includes understanding words, grammar, science, and so on. *Beyond reasonable doubt* implies the probability of P = "He is guilty" must be "large", but "large", thank God, is not defined. That it is not is acknowledgement that not all probability is, and not all probability should be, quantifiable. Besides the idea of "subjective" probability, which in the next chapter I prove is not viable, there is no way to quantify reasonable doubt. But that, of course, would not bar lawyers and judges under the sway of scientism to invent some tedious criteria, perhaps some adaptation of a p-value. Can you imagine trials? "Your honor, my opponent's formula showed a p-value of less than 0.0001, but as you know, the state in Sanity vs. Scientism decided guilt beyond a reasonable doubt must have a p-value smaller than 0.00009. Therefore my client is entitled to be acquitted." The horror, the horror. Jim Franklin's treatise on probability before the advent of reflexive quantification is not to be missed, [80].

4.8 Probability Can Be a Number

Kolmogorov in his 1933s *Foundations of the Theory of Probability* gave us axioms which put probability on a firm mathematical footing. To grasp these, we first need to define an "event", which we can take as a mathematical proposition, i.e. a proposition about something which happens, i.e. is observable, or is about the value of certain numbers. Kolmogorov's first axiom is that the probability of an event, defined over something called an "event space", which we can take loosely as "the set of all those things which can happen", is defined as a non-negative real number. The second axiom is that the probability that something in this event space happens is 1. A third is that this probability is the sum of the probability of each event even if the number of possible events is infinite, but only if the events are "mutually exclusive", "disjoint", or, as classical statisticians would say, "independent."

We're trying to demonstrate that probability can be a number. But the problem with Kolmogorov's axioms is that the first axiom said obliquely but insistently "probability is a number". The second was a repetition, and the third gave a rule for manipulating these numbers. The axioms also require a good dose of mathematical training to comprehend, a situation which contributed to the idea probabilities are always numbers. Anyway, that probability is a number is assumed but not demonstrated by Kolmogorov. Notice, too, that definition of an event space is an assumption of conditionality, albeit hidden, which is why it's usually not recognized.

Kolmogorov's axioms, when stated in their proper mathematical symbology, also embed probability firmly within measure theory, a field in which we can speak of Borel sets, Lebesgue measures, Radon-Nikodym derivatives, and a host of other rich terms. I don't for a second wish to take away from the vast accomplishments of this field, with which any aspiring mathematician must have some familiarity if not intimate knowledge. But investigating these would take us too far from our path. I recommend Patrick Billingsley's delightful book *Probability and Measure*, [22].[3] The level of mathematization of probability at this date is astounding, but we must remember that however some equation was derived, if it is going to be used it must be interpreted in non-mathematical terms.

Different but nevertheless appealing axioms were given by Richard Threlkeld Cox in 1961 [51]. Cox, a physicist, built the foundations of logical probability using Boolean algebra and just two axioms, which are so concise and intuitive that I repeat them here. Their appeal was their statement in plain English and concordance with common sense. Cox's axioms demonstrate probability can be quantified by all finite discrete propositions, and that it can be for many infinite continuous propositions. But there have been complaints, as there always are when dealing with infinity, that certain difficulties can arise if the axioms are pushed too far. Cox's rigor, questioned in e.g. [107, 108] was subsequently stiffened by several authors, e.g. [64, 65], though it's doubtless (given our previous experience with statements about infinity and probability) that these will not be the last word.

Axiom 1: "The probability of an inference on given evidence determines the probability of its contradictory on the same evidence." Axiom 2: "The probability on given evidence that both of two inferences are true is determined by their separate probabilities, one on the given evidence, the other on this evidence with the additional assumption that the first inference is true."

These axioms, conjoined with several other premises about functional equations and the like, yield two fascinating results. First, as was shown above, is that probability is always conditional. Conditionality is assumed in the first axiom. As above, we should never write (say) $\Pr(A)$, which reads "The probability of proposition A", but must write $\Pr(A|B)$ which is "The probability of A given the premise or evidence B." This should come as no shock to logicians, who know that the conclusion of any argument must be conditioned on premises or evidence of some kind. In statistics and other probability texts it is always found that probabilities are written as if they are unconditional, i.e. in the "$\Pr(A)$" form. This mistake is deepened when later in these books "conditional" probability is introduced as if it were a separate thing. Of course, in mathematics it is a burden to write and manipulate things like $\Pr(A|B)$, especially when folks are anxious to do calculations, so the shorthand is forgivable. But it must never be forgotten that mathematical symbols are mere stand-ins, shorthand for ordinary language.

---

[3]Not only was Billingsley a brilliant mathematician, but he was also an accomplished stage actor.

The second interesting result proves probabilities are sometimes numbers. Certainty has probability 1, falsity probability 0, just as expected. And, given some evidence B, the probability of some A plus the probability of the contrary of A must equal 1: that is, it is a certainty (given B) that either A or not-A is true. So we have numbers, but only sort of, because there is no proof that for any A or B, $Pr(A|B)$ will or *must* be a number. And indeed, there can be no proof, as we've seen.

Cox's axioms (and their many variants) are known, or better to say, are only followed by only a minority of physicists and Bayesian statisticians. They are certainly not as popular as Kolmogorov's, even though following Cox's trail can and usually does lead to Kolmogorov. Which is to say, to mathematics, i.e. numbers.

Williamson [225] in his work on mathematically objective Bayes adds to these lists of axioms another, that one (p. 17) "should equivocate between propositions in the absence of evidence." This trick allows everywhere the creation of a numerical probability in the *absence of evidence*. Thus, if all we knew was that a certain proposition was contingent, Williamson's trick would equivocate, which is to say it would add evidence such that the new conditional probability is 1/2. Adding information is unwarranted, of course. Given the bulk of his book, it appears that Williamson's goal is to produce numbers everywhere so that they can be manipulated. But it's not necessary. Consider Walley [217], who develops an entire mathematical framework for manipulating what he calls "imprecise" probabilities, i.e. numeric intervals. Also see [147] on imprecision in probability assessment.

Chapter 5
What Probability Is Not

"Le calcul des chances c'est un calcul des illusions."—Cournot.

Logic is not an ontological property of things. You cannot, for instance, extract a syllogism from the existence of an object; that syllogism is not somehow buried deep in the folds of the object waiting to be measured by some sophisticated apparatus. Logic is the relation between propositions, and these relations are not physical. A building can be twice as high as another building; the "twice" is the relation, but what exists physically are only the two buildings. Probability is also the relation between sets of propositions, so it too cannot be physical. Once propositions are set, the relation between them is also set and is a deducible consequence, i.e. the relation is not subjective, a matter of opinion. Mathematical equations are lifeless creatures; they do not "come alive" until they are interpreted, so that probability cannot be an equation. It is a matter of our understanding.

So-called subjective probability is therefore a fallacy. The most common interpretation of probability, limiting relative frequency, also confuses ontology with epistemology and therefore gives rise to many fallacies. Some authors are keen on not declaring any single definition of probability, and are willing to say that probability changes on demand, but this argument is, as far as I can tell, aesthetical and not logical; see [104]. Probability like truth and (Aristotelian, or "meta") logic is unchanging regardless of application.

5.1 Probability Is Not Physical

Here is perhaps the simplest demonstration that probability is not a physical property. Take the game of craps, played with two six-sided dice. On the come-out, the shooter wins with a 7 or 11. The two-dice total will come to something, possibly this 7 or 11. Is this total caused by "chance" or probability? What we know is that the two-dice total is constrained to be a number between 2 and 12 inclusive.

© Springer International Publishing Switzerland 2016
W. Briggs, *Uncertainty*, DOI 10.1007/978-3-319-39756-6_5

If the *only*—I mean this word in its most literal sense—information that we have is that X = "There will be a game played which will display a number between 2 and 12 inclusive", then we can quantify our uncertainty in this number, which is that each number has probability 1/11 of showing (there are 11 numbers in 2–12, the statistical syllogism provides the rest). If you imagine that this total is made by two dice, then you are using *more* information than is provided. With only X, which is silent on how the total is produced, silent on dice, and silent on many other things, then the probability is 1/11 for every total.

Craps players have more information than X. They know the total can be from 2 to 12, but they also know the various ways how the total can be constructed, e.g. $1+1, 1+2, 2+1, \cdots, 6+6$. There are 36 different ways to get a total using this new information, and since some of the totals are identical, the probability is different for different totals. For example, snake eyes, $1 + 1$, has 1/36 probability.

It should now be clear that these different probabilities are not a property of the dice (or the dice and table and shooter, etc.). If probability were a physical property, then it must be that the total of 2 has 1/11 physical probability *and* 1/36 physical probability! How does it choose between them? Quantum mechanical wave collapses? No, the parsimonious solution is that probability is a state of mind; rather, it states our uncertainty given specific information. Change the information, change the probability.

There have been many attempts to tie probability to physical chance or propensity; for a discussion of both which doesn't quite reach a conclusion, see [16]; for a full-throated defense see [126] and [187]. I think all of these fail. Let A be some apparatus or experimental setup; the things and conditions which work towards producing some effect P, which is, as ever, some proposition of interest. For example A could be that milieu in which a coin is flipped, a milieu which includes the person doing the deed, the characteristics of the coin, the physical environment, and so forth. P = "A head shows." A theory of physical chance might offer $\Pr(P|A) \equiv p$, which is considered a *property* of the system (which I mark with the equivalence relation). Few deny that the argumentative or logical probability of P, perhaps also given A, would have the same value, but the argumentative probability is acknowledged properly as epistemological whereas the equivalence is seen as a physical essence in the same way length or mass might be. Indeed, Lewis's so-called Principal Principle (sometimes also called Miller's Principle) states that the logical probability should equal the physical chance (in my notation; see the definition provided by [115]):

$$\Pr(P|\Pr(P|A) \equiv p) = p. \tag{5.1}$$

The conditions (premises) A are not seen as causative *per se*; they only contribute to the efficient cause of P (or not-P). What *actually* causes P? Well, according to this principle, p: the probability itself. Nobody makes that statement blatantly, but that is what is implied by physical chance. This p does not act alone, it is felt, it is a guiding force, it is a mystical energy. This also is never explicitly stated by it supporters. Physical probability is mysterious, mystical, even. Yet it doesn't operate alone; it requires the catalyst A: A *allows p* to operate as it will, sometimes this way, sometimes that, or sometimes, as in some quantum theories, an infinite number of ways. Just *how p* rises from it hidden depths and operates, how it chooses which cause to invoke—which side of P to be on, so to speak—is a mystery, or, as is again sometimes claimed in quantum theories, there is *no* cause of P *and p is its name.*

The cause of an event cannot be logic, a claim with which any but strict idealists would agree. And since probability is logic, at least sometimes, as chance theorists admit, probability-as-logic cannot be a cause. Therefore, unless there be *no* cause of an event, if we are to save, as we must, the principle of causation, in at least some situations chance itself must be the cause, or contribute to the overall cause. Yet saying chance is a cause is saying Fortuna herself still meddles in human affairs!

As already detailed in Chap. 4, if we knew the initial conditions of a coin flip and knew the forces operating on the coin, we could predict with certainty whether (say) P = "Head" is true. This knowledge is not the cause. Neither are the initial conditions the cause. The equations with which we represent the motion of the spinning object are not the cause; these beautiful things are mere representations of our knowledge of *part* of the cause. No: the coin itself, its handler, and the environment are the material cause; the formal cause is the flip, the physical forces themselves are the efficient cause of the outcome, and final cause is the object of the flip, the Head itself. (I investigate cause more fully in Chap. 7.)

It is obvious enough that because a man does not know the cause of some effect, that it is a fallacy to say the effect has no cause. It follows that because two men do not know the cause, that it is also a fallacy to say the effect has no cause, and so on for all men. Ignorance itself cannot be a cause. Ignorance is the absence of knowledge, a nothingness, and nothing cannot be a cause. Nothing is no thing. Nothing has no power whatsoever: it is *nothing.*

Something caused P to be ontologically true or false. If P were observable, that is; i.e. empirical. Nothing is causing, or *can* cause, Martians to wear hats. This won't be the last time where we observe a decided empirical bias in the treatment of probability. Somehow that bias never was taken up in logic, a subject an enterprizing historian can tackle.

Now if A were a complete description of the coin flip then *p* would be 0 or 1 and no other number. We would *know* whether P was true or false. In everyday coin flips we don't know what the outcome will be; our premises are limited. But some complete A does exist—it would not be proper to call the premises of A "hidden variables"; instead, they are just unknown premises—because some thing or things cause the outcome. And because this is true, it implies that physical chance is real enough, but it only has extreme probabilities (0 or 1). In everyday coin flips we have, or should have, the idea that the conditions change from flip to flip, and that whatever is causing P or not-P is dependent on these changes. Our knowledge of these changes might be minimal or nonexistent, but that means nothing to whatever causes are operating on the coin.

The study of chaos is instructive, which in the simplest form is defined as sensitivity to initial conditions. Causality is not eschewed in chaos theory; indeed, it is often known precisely what the causes are. Equations in chaos theory are fully determinative: if the initial conditions and values of all constants are known—as in the logical *known* and not some colloquial loose more-or-less sense—then the progress of the function is also known precisely. A chaotic equation is thus just like a non-chaotic one, when it comes to knowing what determines what. We know what determines the value f_t for any t if $f_t = f_{t-1} + 1$ and $f_0 = 1$; that which determines it is $f_t = f_{t-1} + 1$. This determination does not disappear because it is, for example, hidden from view and only (say) a picture f_t is shown. Again, because some individual does not know the causes of f_t, or what determines it, does not imply that the causes or determinations do not exist.

Another point needs clarification. You will often hear of a series of data, usually time series, that a model which represents it is wrong because the model does not account for "chaos." We'll later discuss models in much more detail, but this view is incorrect. Any model which fails to predict perfectly would be wrong (predicts non-extreme probabilities) on this view. Like probability, models only encapsulate what we know, not everything that *can* be known. Models, we will learn, do not have to be causal or determinative to provide accurate (in a sense to be defined) forecasts.

Imperfect models are as good as we can get with quantum systems. The strangeness of quantum mechanics has led to two incorrect beliefs. The first is that since we cannot predict with perfect accuracy, no cause exists. The second is that this non-cause which is a cause after all is probability. Here is Stanley Jaki [120, pp. 183–184] quoting Turner: "Every argument that, since change cannot be 'determined' in the sense of 'ascertained' it is therefore not 'determined' in the absolutely different sense of 'caused', is a fallacy of equivocation." Jaki says that this fallacy "has become the very dubious backbone of all claims that epistemology is to be drastically reformulated in terms of quantum mechanics, including its latest refinements of Bell's theorem."

Contrary to what is sometimes read, Bell did not prove causes of QM events do not exist. He only showed that, in certain arrangements, locality must be false. Local efficient causes therefore do not always exist (this assumes we accept all the other standard QM premises). But, for example in a crude summary of a typical QM EPR experiment, if with an entangled pair of particles one is measured to be "spin up" along its "x-axis" while the other instantaneously, and at a great distance away, becomes "spin down", some thing or things still caused these measurements. Locality is violated. What "paradoxes" like this show is that the cause cannot be localized to the places the measurements are taking place. They do not show, and cannot show, *nothing* caused the measurements. Nothing cannot be a cause. *Nothing* is the absence of causes. Some physicists however, e.g. [5, 214], are determined to believe causes aren't real. I'll discuss this view more later.

Bell did not even show that we can never know a QM event's causes. Since this isn't a book on quantum physics, I leave aside the question whether such a proof exists. I tend to think it might, even though the mind of the First Cause is almost surely closed off to us. Meanwhile, QM theory allows us to make good predictions, i.e. specify the probabilities of events. We can say these probabilities "determine" the events in the sense of "ascertain", but we cannot say the probabilities "determines" the outcomes in the sense of "caused." If, after working through the math and fixing the values of some constants, like Planck's—these are our premises—we predict (say) there is an 82 % chance a particle will be in a certain region of space when measured, and we subsequently measure it there, that 82 % itself did not cause the value of the measurement. That 82 % did not cause the probability to "collapse". How could a probability make itself "collapse"?

All we know with QM is that in experiments where we know as many premises as it is possible to know (we think), we can make excellent predictions. But we can do the same in coin flips or dice throws; indeed, casinos make a living doing just that. That systems exhibit stability does not mean that probabilities are causes. Here's another proof. Repeatedly let go of apples from the top of a tall building. We can work out the theory that says the apple will fall with a 100 % chance. It wasn't the 100 % chance that *caused* the apples to fall, it was gravity.

One of latest attempts to avoid admitting defeat about knowledge of QM causes is Everett's Many Worlds (there are other similar attempts). This can be paraphrased as when the wave-function of each of every object which "collapses" (when it "collapses"), it does so across "many worlds", such that each possible value of "collapse" is realized in one (or every one) of these worlds. The number of "worlds" thus required for this theory since the beginning of universe is a number so large that it rivals infinity, especially considering that wave-function equations are typically computed on a continuum. Even if this theory were true, and I frankly think it is not, it doesn't change a thing. Many worlds does not say, and cannot say, why *this* wave-function "collapsed" to *this* value in *this* world. Some cause still must have made it happen *here-and-now*. Many World's is an ontological theory, anyway. Murray Gell-Mann, for instance, offers a purely epistemological view of that theory which better accords with my view, [90]. This section also does not imply the wave-function does not exist, in some sense, as argued in [182]; anyway, the probabilities derived from wave-functions are not the wave-functions themselves, but the probabilities are *conditioned* on the wave-functions, just as all probabilities are conditioned on something. Probabilities derived from wave-functions are just that: derivations and are not the wave-functions themselves, though some (if I understand them correctly) do make this claim. See *inter alia* [84, 85].

5.2 Probability and Essence

A brief note. There is a use of the word "statistics" in physics that roughly means what mathematicians and statisticians think of when considering probability-plus-cause. "Statistics" are found in, or rather are the product of, ensembles, like collections of gas molecules in some closed system, see [32, 168, 215]. This has advanced to a very sophisticated means of making predictions and in gaining physical insight into why these ensembles do what they do. But there is a strange, albeit weak, mysticism here, too. The ensemble itself behaves this way or that, it is said. This isn't strictly true. The gas molecules do behave, and groups of them do this and such. Some thing or things is still causing each molecule to have the orientation and acceleration is has, and so forth. Ensembles are just summaries of these causes, not complete descriptions of them. They cannot be. In this sense, ensembles are like groups (countries, say) of people to demographers. Demographers can make excellent predictions about large groups, but they can't say what the cause of each and every individual to take some action was.

Probability in all its uses, and physics in all of its, cannot escape metaphysics. (No subject can!) Experiments in statistics and physics are grouped, i.e. samples are collected. Why and how? To gain information and because we know, or assume, the items in the collectives (I do not mean to imply von Mises's use of this word) have the same essences, the same natures. This is why we continually return to discussions of cause.

5.3   Probability Is Not Subjective

Cournot was right "*le calcul des chances*" is "*un calcul des illusions.*" Illusions are what allow the false belief that there is a unique probability number for any problem. And this would be all right, except that this has been allowed to develop into a separate theory of probability; i.e. the curious theory of subjective probability. This states that the right probability value, for *any* problem, is that one that "floats your boat" or that gives you warm feelings. It is not surprising that this theory developed in a century when relativism of every kind was in vogue.

Truth and probability are not subjective, *contra* [75, 123, 139, 140, 175], all of whom makes the errors detailed here. The certainty of any proposition is to be adjudged only by the accepted evidence, just like any math problem must be solved only by the conditions set by an examiner. If an instructor said, "Given $x + y = 7$ and $y = -3$" and asks "Solve for x" and a student says, "I *feel*, oh so strongly, that the probability $x = 32$ is 100%" the student would, at least in the old days, be marked wrong. But a subjective probabilist would have to say to the student, "You must be right: your feelings are what count."

When evidence which is not accepted or commonly or tacitly understood is injected into lists of premises, the certainty of the conclusion changes—as it must. This injection is why probability can appear subjective. Yet if probability were subjective then any conclusion would follow from any set of premises. Take "There are 10 Schmenges and 4 Minyks in a room and only one of these persons will come out". We deduce "The probability a Schmenge comes out" as 5/7. But if probability were subjective we could say this probability is "0.01%" or any other number, including 0 or 100%, that gives us a "positive" feeling. Anchoring probability on feelings is a dubious idea: it admits indigestion could play a role in a proposition's chances.

Since probability is not subjective, if the probability of a proposition is different than the one which is deduced (or deducible) from its premises, this *always* implies injected or substituted premises (or a simple mistake in calculation, which is the same thing).

If our evidence is "A heck of a lot of people like this product and Mary is a person", the probability of "Mary likes this product" is not a number because the notion "A heck of a lot of people" is vague and unquantifiable. Yet some might still form a notion of the likelihood of Mary's pleasure based on tacit premises regarding the meaning of "A heck of a lot. . ." These differing premises are what make it appear probability is subjective.

To prove that, let's pick a homelier example. People gamble on sports. One person says the Tigers will win, another says they will lose. That a person gambles does not imply the person has fixed a probability for an event (as we'll see below). But let's suppose each of our two persons do form coherent probabilities for tonight's game. Person A says the probability of victory is 80 % and B says 30 %. This makes probability appear subjective. We ask each how they came to their judgment. A recounts past games, and his comprehensive (he claims) knowledge of tonight's pitchers. B admits the past games, but lays more emphasis on the batting. A discussion like this can range far and wide and last forever. Each party can admit the other's points, but each might stick to, or adjust part way, his assessment. Again, probability seems subjective. But that is only because the premises are so many in number that it is difficult or impossible (and probably impossible) to show how the probabilities that are held are *deduced* from these premises.

And we can't quite tease out whether the stated probabilities are actual deduced probabilities or decisions. They may well be decisions and not probabilities because not all premises imply single-number probabilities, or even numerical probabilities. In those cases, when a probability is stated, it is because a decision has been made or because additional premises have been added or others subtracted.

The key is that if two people agree exactly, precisely, and completely on a set of premises, and on how those premise are probative to the proposition of interest, they *must*, or rather should, agree on the probability. See also [10], who proves the similar specialized proposition "If two people have the same priors, and their posteriors for an event A are common knowledge, then these posteriors are equal." The result is much more general than considering priors and posteriors. Probability is only subjective in the weak sense that the *choice* of premises is not fixed. People are free, in many cases, and especially day to day, in choosing premises. But this freedom is, and should be, greatly reduced in science. The goal of science is to find just those premises which make a proposition as near to certain as possible, so there is no subjectivity.

Another simple problem, as above. Solve for x—give a single, unique number— in the following equation: $x + y = 3$. Of course, it cannot be done: under no rules of mathematics can a unique x be discovered; there are one too many unknowns. Nevertheless, someone holding to the subjective interpretation of probability could tell us, say, "I feel $x = 7$." Or he might say, "The following is my *distribution* for the possible values of x." He'll draw a picture, a curve of probability showing higher and lower chances for each possible x, maybe peaking somewhere near 3 and tailing off for very large and small numbers. He might say his curve is equivalent to one from the standard toolkit, such as the normal. Absurd?

It shouldn't sound absurd. The situation is perfectly delineated. The open premise is that $x = 3 - y$, with a tacit premise that y must be something. The logical probability answer is that there is no probability: not enough information. (We don't even know if y should be a real number!) But why not, a subjectivist might say, take a "maximal ignorance" position, which implies, he assumes, that y can be any number, with none being preferred over any other. This leads to something like a "uniform distribution" over the real line; that being so, x is easily solved for, once for each value of y. Even if we allow the subjectivist free rein, this decision of uniformity is unfortunate because it leads to well known logical absurdities. There cannot be an equal probability for infinite alternatives because the sum of probabilities, no matter how small each of the infinite possibilities is, is always (in the limit) infinity; and indeed this particular uniform "distribution" is called "improper." Giving the non-probability a label restores a level of comfort lost upon realizing the non-probability isn't a probability, but it is a false comfort. Aiding the subjectivist is that the math using improper probabilities sometimes works out, and if the math works out, what's to complain about?

To say we are "maximally ignorant" of y, or to say *anything* else about y (or x), is to add information or invent evidence which is not provided. Adding information that is not present or is not plausibly tacit is to change the problem. If we are allowed to arbitrarily change any problem so that it is more to our liking we shall, naturally, be able to solve these problems more easily. But we are not solving the stated problems. We are answering questions nobody asked.

Subjective probabilists make several errors, sometimes singly and sometimes in concert. These are: to add to or change the premises or evidence, to confuse probabilities with decisions or acts, to assume the propositions which receive the probability must be physical "events," and to assume all probabilities must be numerical.

The first and most blatant is, as was just said, to add to or change the given premises. Gamblers, even intelligent ones who well know the rules of the games they are playing, are notorious for this, saying certain numbers are "due" or that others are "over played." Even if they don't act, they change the accepted premises to accommodate their superstitions about the probabilities. Bayesian statisticians often invent "priors" (which we'll discuss later) to accompany *ad hoc* probability models, these being necessary to solve the equations. But these "objective" "maximum entropy", "ignorance", or "reference" priors are not (or almost always are not) suggested by the given premises.

Now these inventions are more or less harmful, and even at times useful if the premises guessed or invented turn out to match reality, in the sense that the premises which led to assigning probabilities where none was previously possible are agreed on and provide useful decisions. But it is always the case that adding premises changes the problem. Adding any information not tacitly plausible—such as we did when assuming *more* meant *at least half and not all*, or even when we assumed that *all* meant *each and every one without exception*—is to answer a different question. It is not to play the same game. I emphasize this because experience has shown that this point is difficult to accept. People see little trouble adding whatever information they desire to stated, fixed problems merely so that they can arrive at a solution. However practical this is, it is not answering the stated, fixed problem, but answering a new, self-created one. Subjective probability in this sense is like the student on a high school algebra example saying (in the last example), "I think $y = 3$ therefore x must be 0" and that student expecting to be rewarded for his perspicacity. The premise "$x + y = 3$" is *not* equivalent to "$x + y = 3$ and $y = 3$" or "$x + y = 3$ and $y \in [-4, 4]$" or anything else.

Typical justifications for subjective probability, thanks mainly to Bruno De Finetti and Frank Ramsey, involve Dutch books. In the simplest example, there is an "event" of interest. This event is a proposition, usually of the form E = "X will happen at such and such a time and place." We want the probability of E. There are no stated premises, so the problem is not solvable—unless we add premises; in this case, unstateable subjective feelings. Now if our man says that, given his feelings, the probability of E is 0.7, but that given these same feelings the probability of not-E (X will not happen) is 0.5, then our man's emotions have led him astray; his probabilities are said to be *incoherent* because they do not sum to 1. Whatever method of elicitation we use must lead to coherence, or Dutch book can be "made" against him. The man faces sure loss.

His incoherence made plain to him, our man is now invited to think in terms of money or of its mental equivalent, "utility", a fictional currency made of, it appears, discretized emotions. A price for a ticket is set such that our man will pay $1 (or some other amount; however, this figure makes the math easy) if E obtains, and nothing if it doesn't. His opponent, Nature, also given to gambling, gets to choose either side of the bet; that is, Nature can choose to either buy or to sell the ticket, and the man must accept Nature's decision; his only freedom is in setting the price. Obviously, the price must be $0 if the man thinks E impossible, and $1 if he thinks E certain. If the man thinks E impossible and offers a price of (say) $0.4, Nature will decide to sell the ticket. Since, to the man, E will never happen, he will be out $0.4; thus if he truly thinks E impossible, the only reasonable price is $0. The idea is that considering his feelings and knowing Nature gets to decide which side of the bet the man is on, he will set the price which best reflects his idea of E's likelihood. The price (since the payoff is $1, or otherwise suitably normalized) becomes the "probability."

This works as a gamble because if the man's stated probability is not coherent, in the sense noted above, Nature can make a sure profit against the man. For example, if the man's stated probability for E is 0.7 and for not-E is 0.5, Nature could sell the $0.7 ticket for E to happen *and* sell the $0.5 ticket for E not to happen. If E occurs, Nature looses $0.3 on the ticket for E to happen, but it gains $0.5 on the ticket for E not to happen, for a net profit of $0.2. If E does not occur, Nature gains the full $0.7 on the ticket for E to happen, but looses $0.5 on the ticket for E not to happen, again for a profit of $0.2. If the man is incoherent, Nature must necessarily win. Incoherence is a Dutch book, or Nature's arbitrage.

Now none of this tells us whether the man has set his probability for E soundly, even if coherently. Soundness is another matter entirely. But you can now see that setting a probability based on emotional evidence does not give a probability but instead is a *decision*, an act, a bet. Bets are not numerical probabilities, no more than words like *might* and *likely* are numerical probabilities. They are good indicators of propensities, markers of behaviors and proclivities, but they are not themselves probabilities.

These betting justifications for probability are backward: Dutch book theory works because of probability; they do not define probability. Instead of *defining* probability, what they show is that probability is coherent, and that which is not probability might not be. Coherence surely does not guarantee a profit! The empirical bias in probability theory also shows here. There is no betting on the state of the Metalunan interocitor mentioned last chapter, and no Nature to take bets, simply because there are no Metalunans. But there is still probabilities of interocitors taking states.

To amplify that last objection, let Q = "There are exactly 100 Martians and only one wears a hat and George is a Martian." The probability of P = "George wears a hat" given Q is 0.01. But a subjectivist can say, "Based on my utility, it's 83.7%!", or whatever. How can you prove him wrong? There are no experiments that can be run because there are no Martians. There are thus no bets that can be made, because there is no "event" to occur or not. Unless probability is treated as logic, you have nothing to say to the subjectivist and must accept his probability as being right, which is absurd.

5.4 Probability Is Not Limiting Relative Frequency

Probability can be relative frequency, but it makes little sense to speak of limiting relative frequency. If Q = "5 of 10 Martians wear hats and George is a Martian" the probability of P = "George wears a hat" given Q is 1/2, because of the relative frequency of hat-wearing Martians. But Q does not, of course, imply real instances of real Martians, so "relative frequencies" do not have to be ontologically real. The probability of P would also be 1/2 if Q = "5 of 10 Frenchmen in this room wear hats and George is a Frenchman now in this room." Here, Q expresses a relative frequencies of real things. Either way, probability works.

There are other popular ideas of relative frequencies. Von Mises introduced the mathematical idea of a collective, which is an infinite sequence of "attributes" in some set C—and we should stop there. There are no and can be no infinite sequences of any physical thing nor of time. If there were any instances of an infinite number of anything, that thing would be all you would ever see (if indeed you could see anything!) since our finite universe would be filled with that thing. Now whatever mathematical sense this definition makes, and it does make some, it is therefore of no use in measuring the uncertainty of any physical thing. And since most people take an interest in probability and statistics because they want to quantify uncertainty in some real thing, infinite relative frequencies are therefore of no use. Except, possibly, as approximations. We meet these later.

Nor can we use the mathematical apparatus of such a theory on the idea that, even if infinite relative frequencies cannot exist "in real life", we can "imagine" they can. No: we cannot imagine any such thing; we can only say we can imagine it, which is very different. I can imagine a unicorn—mentally picture it, I mean. I can imagine flying through the air or scores of other things. But I can't imagine what an infinite set looks like, or an infinite collective, or an infinite length of time, or Omniscience or Omnipotence. I can speak, think, or imagine analogically about the infinite, but I can never know it.

Limiting relative frequencies are sometimes said to *be* probability, to define them. We take some measurable attribute in an infinite collective, count the number of times the attribute is found—we count in a proper, sophisticated way, of course, doing all this at the limit—and then divide by the total. That limiting relative frequency *becomes* the probability. Thus we have to wait for the long time to know any probability. As Keynes quipped, in the long run we shall all be dead. Limiting relative frequency as a justification of or for a definition of probability suffers from the same flaw as betting as a definition does: they are backward. Study this objection closely. No probability can be known unless the infinite collective be surveyed. Since this never has yet happened, and never will in any of our lifetimes, no probability can ever be known. Probabilities can be made up, of course, in the subjective sense, and this is exactly what frequentists must do whenever they want to make a calculation: make up numbers.

Alan Hájek has done yeoman service in regard to showing the problems with limiting relative frequency with two papers listing 30 arguments against the theory, [105, 106]. These do not exhaust all possible criticisms, nor are all (as he admits) strong, but they are all good and, taken together, are conclusively devastatingly. Let's examine some of these.

Hájek defines hypothetical relative frequentism as: "The probability of an attribute A in a reference class B is *p* [if and only if] the limit of the relative frequency of A's among the B's would be *p* if there were an infinite sequence of B's." Below is my numbering, not Hájek's. I skip some of his more technical criticisms, such as those referring to Carnap's "c-dagger" or to facts about uncountable sets or about different limits for a named sequence, as I think these mix up causality and evidence of the same. I also do not hold with his alternative to frequentism, but that is another matter.

Before we begin, the natural question is why does it seem that frequentism sometimes works? The answer: why does any approximation work? When frequentist methods heed close to the real definition of probability, they behave well, but the farther away they venture, the worse they get. Most "frequentists" implicitly recognize the difficulties of the theory, and tacitly and unthinkingly reject the idea of infinite sequences in practice without realizing that they have kicked over their theoretical support, i.e. that they are not really using frequentism. Here are the biggest objections.

1. In order to know the probability of any proposition, we have to observe an infinite sequence. There are no observed or observable infinite sequences of anything. We can imagine such sequences—we can imagine many things!—but we can never see one. Therefore, we can never know the probability of any proposition. Hájek: "any finite sequence—which is, after all, all we ever see—puts no constraint whatsoever on the limiting relative frequency of some attribute." A finite observed sequence may equal 0.9, but the limit may evince 0.2, or any other number besides 0.9. Who knows?

   In order to picture an infinite sequence, we also, as Hájek emphasizes, must conjure a universe "utterly bizarre" and totally alien to ours. "We are supposed to imagine infinitely many radium atoms: that is, a world in which there is an infinite amount of matter (and not just the 10^{80} or so atoms that populate the actual universe, according to a recent census)." Universes with infinite matter are required if frequentism is to be true; or rather, if any probability is to be had. It's unclear whether Hájek uses *universe* in the philosophical sense of all there is, in which case this criticism has far less force, or in the physical sense of the stuff local to us (or local universe in some set of universes, to speak loosely), in which case the criticism is accurate.

   If you do not see this criticism as damning, you have not understood frequentism. You have said to yourself that "Very large sequences are close enough to infinity." No, they are not. Not if frequentism is to retain its mathematical and philosophical justification. Why? *Every* finite sequence is infinitely far away from infinity. As you'll see, the main critique of frequentism is that it confuses ontology and epistemology, i.e. existence with knowledge of the same.

2. If our premises are E = "This is an n-output machine with just one output labeled * which when activated must show an output, and this is an output before us", the probability of Q = "An * shows" is 1/n as we have been defining it. A frequentist may assert that probability for use in textbook calculations (e.g. which he often does, say, in demonstrating the binomial for multiple throws of hypothetical dice), but in strict accordance with his theory he has made a grievous error. He has to wait for an infinite sequence of activations first before he *knows* any probability. The only way to get started in frequentism is to materialize probability out of thin air, on the basis of no evidence except imagination. Probabilities may be guessed correctly, but never known. Frequentists are thus, whenever they give examples, acting as secret subjectivists.

3. In the absence of an infinite sequence, a finite sequence is often used as a guess of a probability. But notice that this is to accept the argument definition of probability, which in this case is, given only E = "The observed finite relative frequency of A is p" the probability of Q = "This new event is A" is approximately equal to the observed relative frequency p. Notice that logical probability has no difficulty taking finite relative frequencies as evidence.

   For a frequentist to agree, he first has to wait for an infinite sequence of observed-relative-frequencies-as-approximations before he can know the probability that P = Pr(Q|E) is approximately equal to the observed finite relative frequency is high or 1. Nothing short of infinity will do before he can know any approximation is reasonable. Unless he only takes a finite sequence of approximations and uses that as evidence for the probability all finite sequences are good approximations, but then he is stuck in an infinite regress of justifications.

4. Hájek: "we know for any actual sequence of outcomes that they are not initial segments of collectives, since we know that they are not initial segments of infinite sequences—period." This follows from above: even if we accept that infinite collectives exist, how do we know the initial segments of those collectives are well behaved? "It is not as if facts about the collective impose some constraint on the behavior of the actual sequence."

   If hypothetical frequentism is right, to say any sub-sequence (Von Mises's more technical definition relies on infinite sub-sequences embedded in infinite sequences, which is a common method in analysis; here I mean finite sub-sequence) is "like" the infinite collective, is to claim that the infinite collective, which is not yet generated, "reaches back" and causes the probabilities to behave. And this is impossible. In other words, something else here-and-now is causing that sequence to take the values it does, and probability should be a measure of our knowledge of that here-and-now causality.

5. Hájek: "For each infinite sequence that gives rise to a non-trivial limiting relative frequency, there is an infinite subsequence converging in relative frequency to any value you like (indeed, infinitely many such subsequences). And for each subsequence that gives rise to a non-trivial limiting relative frequency, there is a sub-subsequence converging in relative frequency to any value you like (indeed, infinitely many subsubsequences). And so on."

   And how, in our finite existence, do we know which infinite subsequence we are in? Answer: we cannot. We do not. The problem with infinities is anything possible can and will happen. There is no justification whatsoever, if frequentism is true, for treating with any finite sequence.

6. Our evidence is E = "One unique never-before-seen Venusian mlorbid will be built. It has n possible ways of self-destructing once it is activated. It must be activated and must self-destruct. X is one unique way it might self-destruct." The probability of Q = "X is the way this one-of-a-kind mlorbid will self-destruct" is unknown, unclassifiable, and unquantifiable in frequency theory. In logical probability it is $1/n$. Even if we can imagine an infinite collective of mlorbids, there is no way to test the frequency because Venusians build no machines. No sequence can ever be observed.

   Hájek: "Von Mises famously regarded single case probabilities as 'nonsense'..." Yet, of course, all probabilities are for unique or finite sequences of events. David Stove listed this as a key criticism against frequentism. The sequence into which a proposition must be embedded is not unique. Take Q = "Jane Smith wins the next presidency." Into which sequence does this unambiguously belong? All female leaders? All female elected leaders? All male or female leaders elected in Western democracies? All presidential elections of any kind? All leadership elections of any kind? All people named Jane with the title of president? And on and on and on. Plus none of these can possibly belong to an infinite collective. Of course, if probability is logical, each premise naturally leads to a different, not necessarily quantifiable, probability.

7. Hájek: "Consider a man repeatedly throwing darts at a dartboard, who can either hit or miss the bull's eye. As he practices, he gets better; his probability of a hit increases...the joint probability distribution over the outcomes of his throws is poorly modeled by relative frequencies—and the model doesn't get any better if we imagine his sequence of throws continuing infinitely."

   We have to be careful about causality here, but the idea is sound. The proposition is P = "The man hits the bull's eye." What changes each throw is our (really unquantifiable) evidence. The premises for the n-th throw are not the same as for the $n + 1$-th throw. Hájek misses that in his notation, and lapses in the classical language of "independence", which is a distraction. The point is that each throw is necessarily a unique event conditioned on the premise that practice brings improvements. The man can never go back (on these premises) so there is no way to embed any given throw into a unique infinite collective.

8. Our Q = "If the winter of 1941 was mild" to our P = "Hitler would have won the war." A counterfactual. There are many ways of imagining evidence to support P to varying degrees (books have been written!), but there is no relative frequency, not infinite and not even finite. No counterfactual Q-P has any kind of relative frequency, but counterfactuals are surely intelligible and common. A bank manager will say, "If I had made the loan to him, he would have defaulted", a proposition which might be embedded in a finite sequence, but the judgement will have no observations because no loans will have been made. The logical view of probability handles counterfactuals effortlessly.

   Addendum to the mathematically minded, especially in regards to criticisms 1–3. If we assume we know a probability, we can compute how good a finite approximation of that probability is, which is essentially what frequentist practice boils down to. But since, if frequentism is true, we can never know any probabilities, we can never know how good any approximation in practice is.

5.5   Probability Is Not Always a Number Redux

Here is an example adapted from Henry Kyburg [138]. The proposition of interest is Q = "This (really quite tasty) bottle of 2009 Muga Rioja Reserve will break into N pieces when struck by a hammer". The answer is that there is no answer: there is no intrinsic probability of Q. This non-answer answer holds for all Q which are not self-referential, and the reason is that, as said, probability is a measure between propositions. This is why there is no such thing as a probability "of" being struck by lightning, or being bit by a bee, or of dying of a heart attack, or of anything. So much we already know. Subjective probabilists presented with Q and asked for its probability are tempted to provide an answer. But it requires them to provide evidence so that Q can be put into a relation. Such evidence is in the form of a complex proposition, which might look like this E = "I feel this and that." We can now have Pr(Q|E). But is this a number? That all depends on whether E has information about N and about the nature of being hit by a hammer. Note that this would have to be *very specific* information, too, if we are to extract a number. Suppose E = "I've hit bottles before and shards went everywhere." We have the idea from "shards went everywhere" that N will be "large". But since "large" is relative, it can only be large with respect to something, say, sweeping up. An example of very specific information is E = "When hit by hammers bottles like this break into $N = 1, 2, 3$, or 4 pieces." If that doesn't seem realistic (and it doesn't to me), then that only means we have some vague but unstateable idea of what N might be. In other words, the probability of Q isn't a number.

If the *only* evidence you have, received from a question you asked a friend, Q = "How would I know whether that Maduro wrapped *Romeo y Julieta* will taste good?" the probability that P = "This cigar will taste good" is the entire interval from 0 to 1, end points not included. The endpoints are not included because a tacit premise is that the "event" (the observable proposition) is contingent, therefore neither it nor its contrary are logically necessary. If the evidence is instead the tautology T = "This cigar will taste good or it won't", the probability is also the unit interval. Tautologies, or any necessary truth, do not provide information, as we already know, except to possibly tacitly admit contingency. There is no information in Q or T about P other than that P is contingent. Knowing that a proposition is contingent is *some* kind of evidence, but very weak. Contingency tells us that a proposition cannot be a necessary truth. It can, of course, be a local truth.

There is no way to derive a probability of P from either Q or T without adding additional premises. For instance, your friend has uttered Q, but in the back of your mind you recall M = "Maduro wrappers are often good". That is probative of P, and so Pr(P|QM) is not the unit interval, but it is also—and this is key—*not* Pr(P|Q). As above, if we are asked for Pr(P|Q) we are not free to substitute Pr(P|QM).

Now Pr(P|QM) is not the unit interval because a tacit premise of grammar says that *often* means *usually but not always*. To you, it may mean something else. But let's leave it at that. Pr(P|QM) is still not a precise number. It is not a number at all. Contingency is still present, but without it we have a numerically precise definition of *often* no numerical probability is possible. Suppose I insisted O = "*often* means 6 out of 10 times or better", then Pr(P|QMO) $\geq$ 0.6, which is still not a fixed number.

What if we want to bet whether P is true? Perhaps the question is whether to buy the cigar or not, which is very like a bet. Based on Q there is no information whether P. P is not "50/50", so if you say, "Well, I'll take a chance on it", it is not because the probability of P based on Q is 50%. It is because you made a decision, which weighed the pros and cons of buying the cigar, and on balance the pros won. It's hardly even likely that the pros and cons were quantifiable in this mundane decision. Something put you over the edge and caused you to make the purchase, but it wasn't because you relied on some complicated set of decision analysis formulas. It's true that in making this decision you will have considered many premises about the cigar—"It smells good and cigars that smell good typically taste good", "The senorita on the band is pretty and that might make the cigar taste good", etc.—but unless one of these premises was explicitly quantitative, no numerical probability of P will result.

It will now be obvious that most probabilities are in this form. Real quantification is rare; it really only happens in formal problems where quantification is an explicit goal, such as in science or those endeavors in life which seek to approximate science. But, as we'll learn, even in science many of these quantifications are *ad hoc*, and over-certainty is the result. Some of this over-certainty is institutionalized because of the view that all probabilities should be numerical. Numbers are more tangible, they feel "realer" than the vagueness which is reality.

One caution. Some probabilities that at first appear non-numerical or non-precise might be precise. If our evidence is Q = "50 to 70 of the 100 marbles in this bag are cats eyes" then the probability of P = "Pulling out a cats eye" might appear to be 50–70 %, the interval. But it's possible to infer more from the premises. Marbles are discrete, and there could be 50, 51, . . ., 70 cats eyes. The essence of the others do not matter. We could then invoke the statistical syllogism and, through the obvious calculations, come to a probability of 0.6. If there were greater than 50 marbles but as many as 70, then the probability is 0.605, which is notable only because there cannot be, in 100 marbles, any fraction which equals 0.605. As above, we do not need to confirm this by referring to limiting relative frequency of "draws." This *is* the probability. Empirical evidence is not needed for confirmation; indeed, the example is easily changed so no empirical evidence is possible (make them "Martian marbles").

5.6   Confirmation and Paradoxes

Paradoxes, or those claimed to be, crop up from time to time to cast doubt on the ability of probability to provide consistent information. One is Hempel's so-called ravens paradox. This is dealt with adequately in any number of sources, like [115], so that it no longer has any force. Invariably, these paradoxes are resolved after it is discovered some equivocation in wording or that a calculation with impossible values had slid under the radar. Let's do one of these, an example I learned from Deborah Mayo [156] called the "paradox of irrelevant conjunctions."

Loosely quoting Mayo, a hypothesis (proposition) H is confirmed by X (another proposition) in the presence of D if $Pr(H|XD) > Pr(H|D)$ where D is any other proposition. The proposition H is disconfirmed if $Pr(H|XD) < Pr(H|D)$. If $Pr(H|XD) = Pr(H|D)$ then X is irrelevant to H given D. Lastly, H' means "H is false".

Mayo (I change her notation ever-so-slightly) says "a hypothesis H can be confirmed by X, while H' disconfirmed by X, and yet $Pr(H|XD) < Pr(H'|XD)$. In other words, we can have $Pr(H|XD) > Pr(H|D)$ and $Pr(H'|XD) < Pr(H'|D)$ and yet $Pr(H|XD) < Pr(H'|XD)$." In support of this contention, she gives an example due to Popper (again changing the notation) about dice throws. First let D = "A six-sided object which will be tossed and only one side can show and with sides labeled 1, 2, etc., i.e. the standard evidence we have about dice.

Consider the next toss with a homogeneous die, and let H = "6 turns up", H' = "6 does not turn up", X = "An even number turns up." Then

$$\Pr(H|D) = 1/6, \Pr(H'|H) = 5/6, \Pr(H|H) = 1/2.$$

Mayo gives the example due to Popper:

The probability of H is raised by information X, while H' is undermined by X. (It's probability goes from 5/6 to 4/6.) If we identify probability with degree of confirmation, X confirms H and disconfirms H' (i.e., $\Pr(H|XD) > \Pr(H|D)$ and $\Pr(H'|XD) < \Pr(H')$). Yet because $\Pr(H|XD) < \Pr(H'|XD)$, H is less well confirmed given X than is H'. (This happens because $\Pr(H|D)$ is sufficiently low.) So $\Pr(H|XD)$ cannot just be identified with the degree of confirmation that X affords H.

I don't agree with Popper. Because

$$\Pr(H|D) = 1/6 < \Pr(H|XD) = 2/6$$

and

$$\Pr(H'|D) = 5/6 > \Pr(H'|XD) = 4/6.$$

In other words, we started believing in H to the tune of 1/6, but after assuming (or being told) X, then H becomes twice as likely. And we start by believing H' to the tune of 5/6, but after assuming X, this decreases to 4/6, or 20 % lower. Yes, it is still true that H' given X and D is more likely than H, but so what? We just said (in X) that we saw a 2 or 4 or 6: H' is two of these possibilities and H is only one.

"Does X (in the presence of D) confirm H?" is a separate question from "Which (in the presence of X and D) is the more likely, H or H'?" The addition of X to D "confirms" H in the sense that H, given the new information, is now more likely. Mayo recognizes this distinction by quoting Carnap who noted *to confirm* is ambiguous. It can mean (these are my words) "increases the probability of" or it might mean "making it more likely than any other." Pick whichever you like. Neither is a difficulty for probability, which flows perfectly along its course. The problems here are the ambiguities of language and labels, not with logic.

Finally enter the so-called "paradox of irrelevant conjunctions." The idea is if X "confirms" H in the presence of D, then X should also "confirm" HP in the presence of D, where P is some other proposition. There are limits. If P = H', then HP is always false or nonsensical, no matter which X is picked. Ignore these strange cases. As before we can say P is irrelevant to X in the presence of D if $\Pr(X|HD) = \Pr(X|HPD)$. Continuing the example, let P = "My hat is a fedora"; then $\Pr(X|HD) = 1$ and so is $\Pr(X|HPD) = 1$.

The next step in the "paradox" is to note that if X "confirms" H in the first sense above, then $\Pr(X|HD)/\Pr(X|D) > 1$. In our example, this is $1/(1/3)$ which is indeed greater than 1. So we're okay. Now we assume P is irrelevant, so $\Pr(X|HPD) = \Pr(X|HD)$. Divide this by $\Pr(X|D)$, then because $\Pr(X|HD)/\Pr(X|D) > 1$ so too does $\Pr(X|HPD)/\Pr(X|D) > 1$. There are no difficulties so far; just some manipulation of symbols.

Then it is claimed that X, since it "confirmed" H, must also "confirm" HP. Why is this so? Mayo says (still with my notation) there exists an "Entailment condition: If X confirms T, and T entails P, then X confirms P" which is plain enough. "In particular," she says, "if X confirms HP, then X confirms P" by the argument that HP entails P. Here is the magic: "if X confirms H, then X confirms P for any irrelevant P consistent with H. (Assume neither H nor P have probabilities 0 or 1). It follows that if X confirms any H, then X confirms any P."

What has gone wrong? That parenthetical note gives the clue. In our example, H does not entail P, but HP does entail P. What does entail mean? Well, $Pr(P|HP) = 1$. The paradox says X confirms P just because HP entails P. But this can't be right. What's happened here is the conditioning information, which is absolutely required to compute any probability, got lost in the words. We went from "X and HP" to "X and P", which is a mistake. Here is the proof.

If X confirms H, then $Pr(H|XD) > Pr(HP|D)$ (using the weaker sense of "confirmed"). Because P is irrelevant to H and X, then $Pr(X|PD) = Pr(X|D)$ and $Pr(H|PD) = Pr(H|D)$ and $Pr(X|HPD) = Pr(X|XD)$. But if P is confirmed by X, then it must be that $Pr(P|XD) > Pr(P|D)$. But $Pr(P|D)$ doesn't exist: it has no probability. What, after all, does knowing the particulars of some dice have to do with whether or not I wear a fedora? Nothing. Neither therefore does $Pr(P|XD)$ exist. Wearing hats has nothing to do with dice. You can't get there from here. This, after all, is a consequence of P's irrelevancy. (You might tease out P's contingency from the gammar, but objects like $Pr(P|XD)$ are then the unit interval.)

So P can't be confirmed by X in the usual way. What if we add H to the mix, insisting $Pr(P|XHD) > Pr(P|HD)$? Not much is gained, because again neither of those probabilities exist. You can't have inequalities with non-existent quantities. And when we "tack on" irrelevant P, we are always asking questions about $Pr(HP|XD)$ or $Pr(HP|D)$ and not $Pr(P|XD)$ or $Pr(P|D)$.

Result? No paradox, only some confusion over the words. Probability as logic remains unscathed. I presented this argument on my website in 2014 and a semi-anonymous reader "Jonathon D" pointed out the proof of the non-paradoxical nature can be had earlier, by nothing in the very first step there is no probability $Pr(HP|D)$ either. This is true and for the same reasons, but I leave the rest of the argument in place for completeness.

Chapter 6
Chance and Randomness

"What is that chance of that?"—Asked of every statistician by any civilian whenever anything interesting happens.

Randomness is not a thing; neither is chance. Standard statistical interpretation, see e.g. Chapter 1 of [197], assumes randomness is a real physical property. Both randomness and chance are measures of uncertainty and express ignorance of causes and essences. Because randomness and chance are not ontologically real, they cannot cause anything to happen. Immaterial measures of information are never and can never be physically operative. It is *always* a mistake, and the beginning of vast confusion, to say things like "due to chance", "caused by random (chance, spontaneous) mutations", "these results are significant and not due to chance", "no different than chance", "these results are explainable by chance", "random effects", "random variable", "that isn't random", "only random samples count", and the like.

A *coincidence* is a concurrence of observations where one thing is said to be the cause, directly or indirectly, of another thing, but where the cause of the *concurrence* (and not the events) is unknown or immeasurable or suspected to be directed by certain higher powers. The invocation of randomness or chance as this unknown cause is always wrong (but that the higher powers exist might not be). There is an enormous amount of magical thinking which plagues probability and statistics on these questions, including in physics with quantum mechanics and in information theory.

All this holds in quantum mechanics, where the evidence for physical chance appears strongest. What also follows, although it is not at first apparent, is that simulations are not needed. This statement will appear striking and even obviously false, until it is understood that the so-called "randomness" driving simulations is anything but "random". Lastly, how this ties in with information theory and the notion of randomness in that field is given.

© Springer International Publishing Switzerland 2016
W. Briggs, *Uncertainty*, DOI 10.1007/978-3-319-39756-6_6

6.1   Randomness

The English *random* has its roots, so says the *Oxford English Dictionary* [59], from the French, with implications of impetuousness, haste, and violence. It once expressed the range of a piece of ordnance. It wasn't that this ranging distance was variable or chaotic *per se*: *random* was the maximum. One form of *random* in 1624 meant a haphazard route or path. Of course, *(mis)hap* and *hazard* themselves are tied to randomness, *pace* a modern definition: "Having no definite aim or purpose; not sent or guided in a particular direction; made, done, occurring, etc., without method or conscious choice; haphazard." A definition of *random noise* is "unwanted electrical signals caused by randomly occurring transient disturbances...a signal component whose instantaneous amplitudes follow a statistically random or Gaussian distribution." Finally, *random number*, "a number selected from a given set of numbers in such a way that all the numbers in the set have the same chance of selection". Although not from the OED, people will say of observing some quirky event, "That was random."

Random, to us and to science, means *unknown cause*. This view is contrary to many authors who claim, without proof, randomness is a real property and found in, say, (realistically impossible) infinite sets; see [33, 104, 137] among others. *Random* does not and cannot mean *no cause*. Any change (as we shall see much later) must be brought about by something actual, and something actual cannot be "randomness". *Variables*, therefore, cannot be "random"; variables are propositions that take specific values, such as "The temperature will be t", where t is a placeholder for potential values, or is some stated value. Yet some thing or things will cause the eventual t, and this cause or these causes cannot be randomness. *Determine* is a dangerous word. It can mean *caused* or *made known by*. We may know (as we learn next chapter) what determines the truth of a proposition in the sense of what makes the value known, but we may be ignorant of the cause. Randomness is the absence of knowledge of cause or of what determines whether a proposition is true. If you don't know behind which of three doors is the prize, the proposition "It is behind door number 1" is not known to be true because you don't know the cause of the prize being wherever it is and because there is no other information that would let you deduce where the prize is. The outcome is random, even though the prize was put there by some agency.

Coin flips, dice throws, sheeps-knuckle tosses, and the like are caused. But these kinds of events have their own interest. The results are sensitive to their initial and environmental conditions and are therefore chaotic, which as we earlier learned does not mean "not-caused", but they are sensitive to initial (or just-plain) conditions. For some events, it is so difficult to physically manipulate conditions that the event must be ever practically (but not necessarily theoretically) unpredictable. But because these events are as sensitive as they are, tiny, even possibly quantum mechanically sized, deviations in conditions can cause the event to go a certain way or another easily. This is taken advantage of in two ways. Here is one illustration.

So they proposed two, Joseph called Barsabbas, who was also known as Justus, and Matthias. Then they prayed, "You, Lord, who know the hearts of all, show which one of these two you have chosen to take the place in this apostolic ministry from which Judas turned away to go to his own place." Then they gave lots to them, and the lot fell upon Matthias, and he was counted with the eleven apostles. *Acts* 1, 23–26.

It is here that agency might enter the story, as it often does when speaking of randomness. The apostles reasoned in one of two ways. The first is that they trusted that God would "tweak" the conditions of the tumbling lots so that they would land in the optimal way, in the sense of selecting the optimal apostle. I have seen Buddhists at temples in Taiwan, for instance, do a similar thing with crescent-shaped blocks of wood called *bwa bwei*. These are a pair of asymmetric, hand-sized blocks which are thrown onto stone or dirt floors; they bounce around a bit, and come to rest with one or both *bwa bwei* having the round or flat size uppermost. Questions are asked of the "device" and answered depending whether the sides match or mismatch. The appeal is to a higher power, but one which is somehow unwilling to perform a macro feat, as God could easily do, for instance in the case of the lots, by having had the apostles places the lots on the ground and then God could turn them so that they pointed to Matthias. The same is true of *bwa bwei*. The faithful could merely place the *bwa bwei* on the ground and ask the local deity to move them to the position which matches the correct answer. But in both cases this seems like asking for a great amount of work from God or the deity. Instead, when asking for interventions, we ask for the smallest possible assistance, the tiniest adjustment to the conditions, that which requires an almost infinitesimal physical force, so that the device is caused to take its eventual state in such a way that the higher power is not unduly taxed. This act on our part recognizes the sensitive and even precarious nature of the device; indeed, it makes active use of it.

But there is a second sense in which we can interpret the choosing of the substitute apostle which is vastly more plausible. This sense won't work for the *bwa bwei* petitioner, who simply is asking for a *physical* intervention. Because the apostles understood that the tossing of the lots is unpredictable and nearly impossible to gaff, i.e. to finagle or scam, there would be no sense of *human* agency in the choice of the next apostle. If the eleven would have had a vote, Joseph might have won or Matthias would have. There would have been some apostles in favor of Joseph, and some in favor of Matthias. A discussion would begin and politics would enter. And people have long memories. Feelings could be hurt. Since both men were eligible, why not let some unpredictable device make the selection so that *everybody* is excused from making a choice? This is, after all, why we let referees decide who gets the ball first by coin flips (I have more to say about this below). Randomness, i.e. unpredictableness, solves some political conundrums.

What makes the first example different from the statistician waiting to see what value a "random variable" takes? Only this: the higher power is not usually thought to be a wilful agency, and is instead some vague, almost mystical hidden power. One example is "noise", the "error" or "residual" or "ϵ" term in a regression or in a simulation (about which, more below). The value of some ϵ is thought to come about "randomly", and if this "randomly" is thought about at all, and often it is not, it is often thought "random" mystical forces are performing the cause. Hence statisticians will talk about "sampling" so that these mysterious forces "cancel" each other out upon repeated "trials". It is often said that "probability distributions" underlie a set of observations, which again imply probability is cause. The main exception to this magical thinking is electronic engineering and the like, where engineers are forced to think about causes of everything that happens, though even in these fields, thinking that randomness is a cause is not unknown. Opposite this are those uses of statistics applied to human behavior, where what causes the "ϵs" is always said to be randomness. Whatever causes any ϵ to take the value it does, it is not randomness.

I either have in my pocket as I write this my pipe or I don't. That is, I own a pipe and sometimes smoke it while writing, except when I need both hands I put it in my pocket, or I don't carry it at all hence it can't be in my pocket. The proposition of interest is P = "Briggs had a pipe in his pocket when he wrote this proposition." P is *random* to you, because the only evidence you have is that which I provided, which is not sufficient for you to form a unique probability. Of course, you can always add evidence which is not provided, but by that maneuver you make probability subject to whim, which is to say subjective. P is not random to me, because I possess enough extra information that the P is an extreme probability, either 0 or 1.

That is, you must judge Pr(P|Briggs owns a pipe and...) whereas I must judge Pr(P|I have my pipe). The former probability is not a fixed number (it may be the unit interval *sans* endpoints if you consider the tacit premise that the event is contingent; that "I do or don't have a pipe" is a tautology and provides zero information), but the latter probability is 1 (and would have been 0 if I changed by evidence to "I don't have my pipe").

Randomness therefore exists when the probability of a proposition given stated evidence or model is not 0 or 1. That is, randomness applies to the premises (or model) we have and *not* the outcome. All uncertain events are thus *random*. An event is *random* only if it is unknown (in its totality). A state is *random* if it is unknown. *Randomness* is thus a synonym for *unknown*. That, and nothing more.

Statisticians speak, somewhat incorrectly as we have just seen, of *random variables*. These are mathematical creatures, propositions which contain or represent an unknown quantity. For example, S = "Sally's grade point average is x" where x is unknown, i.e. "random." S is neither true nor false—it does not have an extreme probability—and can be neither true nor false because there is no premise with which to judge it, except perhaps that "The grade point average will be some number in this set". But even given that evidence, the proposition has no probability because x is not a number but a placeholder. It's like saying "The color is _____". It is an incomplete statement. This seemingly trivial point is crucial to retain. There is no observation of x. Once we do observe an x, the proposition becomes true with respect to that observation. Thus *random variable* means a proposition with an unknown quantity (the quantity may of course be multidimensional).

Of course, there is a purely mathematical way to speak of "random" variables, i.e. as some kind of measurable function from a probability space into a measure or state space, and so forth. However useful this technique is for computation, and it is, when applying probability to real propositions of interest in arguments, we must not forget that the mathematics are not real. I speak more on the Deadly Sin of Reification which arises from attempts to give equations life in the discussions of modeling.

Chance is identical to randomness in most senses, though it often comes with connotations of unpredictability. Take a "game of chance" such as craps, which is based around a two-dice total, or score. The bounds of the total are deduced from the rules of the game. These bounds are, as is obvious, predictable, so chance does not mean complete inability to predict. There are any number of physical mechanisms that cause each dice total, causes of which we are mostly or completely ignorant. We know the causes must be there, we just don't know what they are for individual plays. We do know there are many causes: imagine the bouncing rolling dice flopping around, buffeted by this and that. If we knew some of these causes for individual rolls—perhaps we could measure them in some way as the dice fly; say, by noting the walls of the table are cushier and more absorbent than usual—then we could incorporate that causal information and use this to update the probabilities of the totals. A 7, which is a winning score on the come out, might be more or less probable depending on how the information "plays". The probability changes because the information changes. Incidentally, unless your knowledge of cause is complete, you might not necessarily beat the casino for any single game, but if you have good causal knowledge, you will beat them over multiple games. It is for this reason that casinos ban contrivances that could measure causes or proxies of causes. In any case, chance is unpredictability, which is a synonym of ignorance, which is what random means.

6.2 Not a Cause

Randomness is not a cause. Neither is chance. It is *always* a mistake to say things like "explainable by chance", "random change", "the differences are random", "unlikely to be due to chance", "due to chance", "sampling error", and so forth. Mutations in biology are said to be "random"; quantum events are called "random"; variables are "random", and all of these things take values be*cause* of chance. An entire theory in statistics is built around the erroneous idea that chance is a cause. This theory has resulted in much heartbreak, as we shall see.

Flip a coin. Many things caused that coin to come up heads or tails. The initial impetus, the strength of the gravitational field, the amount of spin, and so on, as we have discussed previously. If we knew these causes in advance, we could deduce—predict with certainty—the outcome. This isn't in the least controversial. We *know* these causes exists; yet because we might not know them for *this flip* does not imbue the coin with any magical properties. The state of our mind does not effect the coin in any, say, psychokinetic sense.

Pick up a pencil and let it go mid air. What happened? It fell, because why? Because of gravity, we say, a cause with which we are all familiar. But the earth's gravity isn't the only force operating on the pencil; just the predominant one. We don't consider the pencil falling to be "random" because we know the nature or essence of the cause and *deduce* the consequences. We need to speak more of what makes a causal versus probabilistic model, but a man standing in the middle of a field flipping a coin is thinking more probabilistically than the man dropping a pencil. Probabilities become substitutes for *knowledge of* causes, they do not become causes themselves.

The language of statistical "hypothesis testing" (in either its frequentist or Bayesian flavor with posteriors or Bayes factors; for the latter, see Chapter 4 in [100]) is very often used in a causal sense even though this is not the intent of those theories. We must acknowledge that the vast majority of users of models of uncertainty think of them in causal terms, mistakenly attributing causes to variously *ad hoc* hypotheses or to "chance." Attributing anything to "chance" is to attribute it to a chimera, a ghost. This kind of attribution language, which is nearly universal, also implies that the parts not attributable to "chance" are attributable to the other "variables" in the model in a causal sense, which is an unjustifiable stance, as we'll see next chapter.

Specific examples will be offered later, but for now suppose the user of a model of income has input race into that model, which occurs in two flavors, J and K (these letters are next to each other on my keyboard). The "null" hypothesis will be incorrectly stated as "there is no difference" between the races. We know this is false because if there were no difference between the races, we could not be able to discern the race of any individual. But maybe the user means "no difference in income" between the races. This is also likely false, because any measurement will almost surely show differences: the measured incomes of those of race J will not identically match the measured incomes of those of race K. Likewise, non-trivial functions of the income, like mean or median, between the races will also differ.

If the observed differences are small, in a sense to be explained in a moment, the "null" has been failed to be rejected; it is never accepted. Why this curious and baffling language is used is explained in the Causality chapter when we discuss falsifiability. For now, all we need know is that small (but actual differences) in income will cause the "null" to be accepted. (Nobody really thinks in terms of failing to reject, despite what the theory says.) When the "null" is accepted it is repeated that there is "no" difference between the races, or that any differences we do see are "due to", i.e. *caused by*, chance.

But chance isn't a cause. Chance isn't a thing. There is no chance present in physical objects: it cannot be extracted nor measured. It cannot be created; it cannot be destroyed. It isn't an entity. The only possible meaning "due to chance" or "caused by chance" could have is magical, where the exact definition is allowed to vary from person to person, depending on their fancy.

Some thing or things caused each person measured to have the income he did. Race could have been one of these causes. An employer might have looked at an employee and said to himself, "This employee is of race K; therefore I shall increase his salary 3% over the salary I would have offered a member of race J." Or he might not have said it, but did it anyway, unthinkingly. Race here is a partial cause.

The man did not receive his entire salary (I suppose) because he was of race K. This kind of partial cause might have happened to some, none, or all of the people measured. If the researcher is truly interested in this partial cause, then he would be better served to interview whoever it is that assigns salaries and so discover the causes of salary in each case. Assuming nobody lies or misremembers and can bring themselves to proper introspections—an assumption of enormous heftiness—this is the *only* way to assign causes. But actual measurements are time consuming and expensive and, if employed universally, would slow research down to a crawl. Results must needs be had! The advantage to taking this more measured pace would be that many results wouldn't be absurd, like the results of many studies conducted with ordinary statistics surely are. Why? Researchers have been falsely taught that if certain statistical thresholds are crossed, causality is present. This fallacy is the cause of the harm spoken of above.

Even if the null is not rejected it is still possible that some or even all of the people measured had salaries in part assigned because of their race. There isn't any way to tell looking *only* at the measured incomes and races. If the null is accepted, *no* person, it is believed, could have had their incomes caused partially by their race. Again, there isn't any way to tell by looking only at the data. But when the null is accepted, almost all researchers will say that causality due to race is absent—replaced, impossibly, by chance; *or* the researcher bent of proving differences in cause will say, if the null is accepted, that the differences are still there, he just can't now prove it. The truth is we have no idea and can have no idea, looking *just at measured race and income and at nothing else* why *anybody* got the salaries they did. To say we can is wild invention, to replace reality with wish.

On that latter point, when the null is accepted, but the researcher had rather not accept it, perhaps because his hypothesis was consonant with his well being or it was friendly to some pre-conception, he immediately reaches to factors outside the measured data. "Well, I accepted the null, but you have to consider this was a population of new hires." That may be the case, but since that evidence did not form part of the premises of the model, it is irrelevant *if we want to judge the situation based on the output of the model.* I have much more to say on this when discussing models. It is anyway obvious, that, to his credit, the researcher is looking for causes. Even if he gets them wrong, that is always the goal, or should be.

Another popular fallacy, when "nulls" are rejected, is the I-can't-think-of-another-reason-so-my-explanation-is-correct fallacy. If the classical (or any) procedure says there are (which we could have known just by looking) differences, then the researcher will say those differences are caused by the differences in race. He will assume his cause always applies, or at least it mostly or usually applies. Yet he never will have measured any cause, so he is being boastful, especially considering how easy it is to reject "nulls". Later when discussing hypothesis tests, I cover this false dichotomy in more details.

Aristotle (*2 Physics v*) gives this example of what people mean when they say "caused by chance":

> Some people even question whether [chance and spontaneity] are real or not. They say that nothing happens by chance, but that everything which we ascribe to chance or spontaneity has some definite cause, e.g. coming "by chance" into the market and finding there a man whom one wanted but did not expect to meet is due to one's wish to go and buy in the market...
>
> A man is engaged in collecting subscriptions for a feast. He would have gone to such and such a place for the purpose of getting the money, if he had known. [But he] actually went there for another purpose and it was only incidentally that he got his money by going there; and this was not due to the fact that he went there as a rule or necessarily, nor is the end effected (getting the money) a cause present in himself – it belongs to the class of things that are intentional and the result of intelligent deliberation. It is when these conditions are satisfied that the man is said to have gone "by chance". If he had gone of deliberate purpose and for the sake of this—if he always or normally went there when he was collecting payments—he would not be said to have gone 'by chance'.

Notice that chance here is not an ontological (material) thing or force, but a description or a statement of our understanding (of a cause). Aristotle concludes, "It is clear then that chance is an incidental cause in the sphere of those actions for the sake of something which involve purpose. Intelligent reflection, then, and chance are in the same sphere, for purpose implies intelligent reflection." And "Things do, in a way, occur by chance, for they occur incidentally and chance is an incidental cause. But *strictly it is not the cause—without qualification—of anything*; for instance, a house-builder is the cause of a house; incidentally, a flute player may be so". Chance used this way is like the way we use *coincidence*.

There is also spontaneity, which is similar: "The stone that struck the man did not fall for the purpose of striking him; therefore it fell spontaneously, because it might have fallen by the action of an agent and for the purpose of striking." But this does not mean that nothing caused the stone to fall. It could very well be that the stone was made to fall by some wilful agency, as many might imagine, but because we have no evidence of this, save our suspicions, we can't be sure. We can have faith that the stone was sent by God for some purpose, we can have superstition that some evil entity caused the tumble, we can believe it was just "one of those things". Which is the right attitude, faith, superstition, disbelief? It can't be known from the concurrence alone. Just like the in the race-income example, we have to look outside the "data". This necessity is ever present. Data alone are meaningless.

Persi Diaconis and Fred Mosteller provide a well known definition of *coincidence*: "A coincidence is a surprising concurrence of events, perceived as meaningfully related, with no apparent causal connection," [58]. The phrase "no apparent causal connection" is apt but incomplete, as we now see. Coincidences are rather taken to prove causation, by whom or what we might not know but only suspect.

Lastly, again Aristotle, "Now since nothing which is incidental is prior to what is *per se*, it is clear that no incidental cause can be prior to a cause *per se*. Spontaneity and chance, therefore, are posterior to intelligence and nature. Hence, however true it may be that the heavens are due to spontaneity, it will still be true that intelligence and nature will be prior causes of this and of many things in it besides." In other words, "posterior to intelligence and nature" means they come after as explanations and not prior as causes (Bayesians ought to take pleasure in that choice of words).

The language of calling chance, randomness, and spontaneity explanations is risky because *explanation* quickly becomes *caused*, and, as just said, we can't know the "higher" cause of any event just by examining the data at hand. It is thus better to avoid the words altogether, especially in science where the goal is to understand cause, unless one can be exceedingly careful.

6.3 Experimental Design and Randomization

A purposely absurd, yet telling, example. You're a statistician and new recruit to ISIS assigned to crucify three score perceived enemies. However, you've run out of wooden crosses. But there are some sturdy metal poles that you think might make good substitutes. So you go to the chief and say, "Boss, I want to prove that crucifixion by metal pole is as efficacious as by wooden cross. I have drawn up an experimental design to randomize victims to either wood or metal. If all goes well, I'll be able to show that death by metal-pole crucifixion is statistically identical with wooden-cross crucifixion."

Why is the example absurd? It is *true* that since metal-pole crucifixion hasn't been tried, victims *might* not die. The event is contingent. And since statistical evidence in the form of a "gold standard" randomized controlled trial hasn't been supplied, how could anybody believe metal works as well as wood? Don't skip lightly over these questions. Their answer explains why randomization isn't needed, and that some experiments are of no utility. Essence and cause are once again present.

What is the purpose of an experiment? To provide evidence probative towards some proposition of interest. Here the proposition is, "Victims will die by metal crucifixion". Evidently, the "data" that can be gathered is probative. If nobody dies while welded (tied, trussed, nailed, or whatever) to a metal pole, then we have learned that metal-pole crucifixion does not work, or does not work well. Contrariwise, if everybody dies strapped to a metal pole, this is also probative, and we'll have evidence giving weight to our proposition. Since the experiment obviously fulfills the desideratum of an scientific experiment, we haven't discovered why the example is absurd.

Similarly, randomization is meant to guard against the possibility of experimental error of a certain kind. Everybody "knows" randomization is a good thing; indeed, it is believed essential for a quality study. But since your experiment will use randomization, again we have fulfilled the standard desideratum of experimental design. And we still haven't discovered why the example is absurd.

Could it be because, "It's *obvious* that metal poles are no different than wooden crosses"? This is necessarily false. If there were no difference, they would both be made of the same material; and we'd probably not have different words nor would we be able to form separate mental images for the two objects. They are not the same; they are different; *of course* they are different! Plus, this difference was *acknowledged* by the design of the experiment. If you had thought they were identical, there would have been no need to gather new evidence. It is because there are *known* differences that you proceeded. Still no absurdity.

The answer is this: it's *obvious* that people crucified to metal poles will die as they do when tied to wood crosses because the *cause* of death is excruciating exposure. This is an induction, and a true one, coming to us in syllogistic form as we saw in Chap. 3. We induce that the *essence* of the kind of death, the experimental "outcome", is excruciating exposure and that however one is strung up, be it metal, plastic, wood, or some other substance, the result will be the same. We already have the evidence we need that makes the proposition of interest true, evidence supplied by induction-argument. Notice also that our interest was always the *cause* of death and nothing to do with metal poles *per se*. We didn't want to gather evidence that would make the proposition of interest likely, we wanted to know what *caused* it to be true. And this we got for free.

Randomization wasn't needed. But perhaps that's because the experiment itself wasn't needed. Perhaps in other instances the blessings provided by randomization *are* needed. They aren't, as we shall see. But before I can prove that, we need to understand more the purpose of experimentation.

The difficulty lay in the definition of the experiment. An experiment is the process of discovering information probative to a proposition of interest. Experiments can be active or passive. They are passive when nothing but mental labor is involved in discerning this evidence as was the case in the crucifixion example, or they are mixed passive and active in cases where data is gathered in some (usually mechanical) fashion. So-called observational experiments are more passive than active, but they can be just as active as so-called controlled experiments; e.g. "chart reviews" in medicine. None of these dividing lines are sharp, and most experiments are really mixtures of these types, but controlled experiments usually see evidence *generated* or *caused newly to be made* under conditions *controlled*, to various extent, by the experimenter.

The experimenter in *any* kind of experiment is the person responsible for the three most important things: deciding the proposition of interest, stating what evidence is probative, and then gathering it. This isn't an empty statement. Propositions of interest are not free for the asking. They are related almost always to decisions people want to make in the face of uncertainty. The possibility of mixup is great, because propositions that are answerable are often stand-ins or proxies for what is of real interest. Also, readers of experiments often mix up or misunderstand just what the proposition of interest was that guided the experimenter. By the time most "studies" reach the press they are as badly garbled as a Shakespearean sonnet conveyed by the Telephone Line game played by first graders.

Propositions of interest are anything from "The weight of this elementary particle is y" to, "The value of this biological measure is w", to "The amount a person will spend is z", etc. The times when we can deduce or induce evidence which tells the cause of propositions of interest are rare. Or, rather, they only seem so because nobody records the "mental experiments" like the crucifixion example. But because they're not recorded, they come to seem as if they are not experiments, which is too bad. We only come to know the "hard cases", i.e. those propositions where the evidence of cause is inconclusive. Or we know those cases in which the proposition of interest has been deduced *given* a set of circumstances (premises), and the experiments are such that they verify the conditions (premises)-proposition concordances. These are, of course, local deductions based on contingent premises and not necessary truths. If they were necessary truths, we'd again have no need to perform any experiment (except for pedagogical purposes).

What evidence is probative? This is the real question. Let's work with an example. Y = "The value of this biological measure is *y*". If I claim X = "This biological measure can only be *y* = 120 mm/Hg", then the probability Y takes any value but 120 mm/Hg is 0. Or I could have said, X = "This measure can only be 120 mm/Hg or 160 mm/Hg", then the probability that it is 120 m/Hg is 0.5 and so forth. But where do these X come from? I made them up. Given the X I supplied, the probability of Y is *deduced* by the rules of probability (which only takes as condition the information supplied *and none other*). But if my audience is genuinely interested in Y, they are unlikely to be convinced that my probative X—and it îs probative—is proper. What experiments are looking for, then, are X which are themselves true or can be reasonably believed given another or "outside" our observed set of premises.

For instance, suppose I conducted an experiment to take actual measures of this Y, and further suppose that each time the measurement was 120 mm/Hg. I announce, given my experiment—my X, which are my measurements—the probability of Y = "The measure is 120 mm/Hg" is high. Now you can choose to believe this or not. If you do, you are supplying tacit premises of the form, W = "This Briggs is honest and his measurements were error free and in the milieu, form, and type that I expect." Then, given W and X (the conjunction), the probability is Y is high. But if you reject W and suppose instead, "I haven't a clue what Briggs is on about; why are all the numbers the same?; maybe they weren't of the form I expected", then the probability of W and X is not high. (The probability of Y given X is, no matter what, high.)

Real experiments tighten this. They list all the premises which led to the measurements or collection of data thought probative of Y. No matter how good a job I do at listing premises (explaining the experiment), you still must trust if you are to believe. The best experiments find those premises where the probability of Y is extreme or high and where trust (or faith!) is high, and the worst experiments find premises which are murkily related to Y and were trust is low. The premises in which the probability of Y is extreme or high are those most related to the cause or causes of Y.

Control, true control, is what produces the best evidence, not randomization. To understand what causes some thing to happen, the ideal experiment is of course that which focuses on that thing or things that are the cause. If we can hold fast every condition which we assume might be a cause of Y (our proposition of interest) and vary or manipulate only one, we are then certain that the changes in Y are caused by this manipulated condition. This is a local truth. It is not a necessary truth because it presumes that we have identified all possible causes. Since Y will usually be contingent, it is likely we might err in this presumption, especially if Y concerns complicated matters like human behavior. Of course, holding *all* things constant is a tremendous demand, one that perhaps can never be met in practice. This is the implication of, for example, Nancy Cartwright's work, [39], who has repeatedly emphasized the necessity of identifying true causes.

If we *can* hold everything constant and manipulate one X and witness the changes in Y, then we can make statements like this: "Assuming all other things constant, and set at these certain levels, when X $= x$, Y $= y$." This produces a local truth that Y $= y$ (and possibly even a necessary one if the "all things constant" is sufficiently tight, i.e. deduced from axioms). Probability is not needed: there is no uncertainty. This easily extends to multidimensional X and Y. Since this is so, we don't need randomization when we can control. Indeed, randomization is the *opposite* of what we want. Randomization could introduce *variation* into those things which are potentially causative and we thought we were controlling!

True controlled experiments demonstrate cause; they confirm cause. But because controlling everything that might be a cause of Y is difficult or practically impossible when Y has many causes, as does human behavior, we often have to settle for experiments where only some, or maybe even no, things that are causative can be controlled or observed. We posit race as one of many causes of income. We can measure race and income, but it is clear that race is not (or almost never not) the sole cause. We can now only say things like this: "Assuming race is causally related to income, and given some observed race-income pairs, as well as assuming some technical modeling details, when race is K, the probability of incomes (for people we haven't yet measured) larger than x is p_K, whereas if race is J, the probability is p_J." We'll discuss "causally related" in the next chapter. We would also have to say "Assuming race is *not* causality related to income, and given some observed race-income pairs, as well as assuming some technical modeling details, when race is J *or* K, the probability of incomes (for people we haven't yet measured) larger than x is p; i.e. p is the same for J and K."

Causal relations are what drive the probability. The uncertainty is only in those causes which we could not measure. And if we cannot measure potential causes because we don't know what they are, then randomizing does nothing for us. It gives probability no special boost; randomizing does not, as it is tacitly thought, bless statistical results. If we don't know what all potential causes are in some human experiment, for instance, we then do not know if person 1 has a potential cause and person 2 does not and so on, therefore randomizing does nothing. And there may be *many* causes each person possess that are unknown to us, thus mixing people up helter-skelter is absolutely no guarantee of producing equal mixtures of people with these unknown causes in each of our experimental groups. We are flying blind. What probability does for us is to regain partial, but still hazy vision for those causes we have assumed are operative.

Besides, "randomizing" isn't even a thing that one can do in the mystical sense people usually take that word. Since randomness only means unknown, *to randomize* can only mean *to make unknown* and that is the opposite goal of any experiment! Adding "randomness" to experiments does not make them valid, it makes them worse. Thus *randomizing*, if it means anything, means *removing control*. And that is fascinating.

There is still the vague unsettling notion that "randomizing" does *something* for us. And this is true. It can, in some situations, restores trust, or rather allay suspicions of deceit. But this is only when randomizing is used in its proper sense of removing control. When dealing with duplicitous lying unscrupulous scheming self-deceping conniving canny chiseling human beings we need often to have some procedure to lessen the chance of falsehood. This is why we have referees flip a coin to decide who gets the ball first. The coin flip is random in the sense of unknown. It is also chaotic in the sense that it is sensitive to conditions. We all understand it is difficult—but not impossible!—to manipulate the flip so that a given outcome occurs. In this way, we let "nature" decide who gets the ball. Or, better, we remove the knowledge that some human being is cheating us in some way. Feelings aren't hurt. We also remove control (randomize) patients in medical trials. Doctors are as prone (and maybe more?) to self-deception as the rest of us, and they can too easily veer patients into treatment and control groups to make the treatment seem better or worse than it really is. So we remove control from the doctor and give it another, say, a statistician. This isn't ideal because usually the statistician thinks he is "randomizing" in the mystical sense. It would be far better to examine each patient and control assiduously which group in an experiment this patients goes, and for this control to be open so that fears of abuse are minimized. Stephen Senn has many insights on this subject, [194, 196]. And notice physicists don't rush to "randomize": they control. That's because cheating, including self-cheating, is far less of a problem (but not non-existent).

6.4 Nothing Is Distributed

People sometimes speak as if random variables "behave" in a certain way, as if they have a life of their own. Thus "X is normally distributed", "W follows a gamma", "The underlying distribution behind y is binomial", and so on. To behave is to act, to be caused, to react. Somehow, it is thought, these distributions are *causes*. This is the Deadly Sin of Reification, perhaps caused by the beauty of the mathematics where, due to some mental abstraction, the equations undergo biogenesis. The behavior of these "random" creatures is expressed in language about "distributions." We hear, "Many things are normally (gamma, Weibull, etc. etc.) distributed", "Height is normally distributed", "Y is binomial", "Independent, identically distributed random variables".

I have seen someone write things like "Here is how a normal distribution is created by random chance". Wolfram MathWorld [219] writes, "A statistical distribution in which the variates occur with probabilities asymptotically matching their 'true' underlying statistical distribution is said to be random." There is no such a thing as a "true" distribution in any ontological sense. Examples abound. The temptation here is magical thinking. Strictly and without qualification, to say a thing is "distributed as" is to assume murky causes are at work, pushing variables this way and that knowing they are "part of" some mathematician's probability distribution.

To say "X is normal" is to ascribe to X, or to something, a power to *be* "normal" (or "uniform" or whatever). It is to say that forces exist which *cause* X to be "normal," that X somehow *knows* the values it can take and with what frequency. If this curious power notices we have latterly had too many small X, it will start forcing large ones so that the collective exhibits the proper behavior. This is akin to the frequentist errors we earlier studied.

To say a thing "has" a distribution is false. The only thing we are privileged to say is things like this: "Give this-and-such set of premises, the probability X takes this value equals that", where "that" is calculated via a probability implied by the premises. (Ignore that the probability X takes *any* value for continuous distributions is *always* 0; this is discussed much later under measurement.) Probability is a matter of ascribable or quantifiable uncertainty, a logical relation between accepted premises and some specified proposition, and nothing more.

Observables also do not "have" means. Nor do they have variances, autocorrelations, partial or otherwise, nor moments; nor do they have any other statistical characteristic you care to name. Means and all the rest can be calculated *of* observables, of course, but the observables themselves do not possess in any metaphysical sense these characteristics. This goes for observables of all kinds. Time series are, in some analyses, supposed to be "stationary". A stationary process, it is said, has the property that the mean, variance and autocorrelation structure do not change over time. Actual functions of observables such as means do change over time, as all know. Premises from which we deduce probabilities if they include observable propositions can also change, and thus so can the probabilities. Specific model premises which hold fixed various parameters (about which much more later) can be assumed or not. That is all stationarity means epistemologically. Causes of observables can and often do change, but since probability is never a cause, neither can stationarity nor any other statistical characteristic be a cause.

Back to Sally and her grade point. We had S = "Sally's grade point average is x". Suppose we have the premise G = "The grade point average will be some number in this set", where the set is specified. Given our knowledge that people take only a finite number of classes and are graded on a numeric scale, this set will be some discrete finite collection of numbers from, say, 0 to 4; the number of members of this set will be some finite integer n. Call the numbers of this set $g_1, g_2, \ldots, g_n$.

As said above, the probability of S given G does not exist. This is because x is not a number; it is a mere placeholder, an indication of where to put the number once we have one in mind. It is at this point the mistake is usually made of saying x *has* some "distribution", usually normal or perhaps uniform (nearly all researchers I have seen in applications of GPA say normal). They will say "x is normally distributed." Now if this is shorthand for "The uncertainty I have in the value of x is quantified by a normal distribution", the shorthand is sensible—but unwarranted. There are no premises which allow us to deduce this conclusion. The conclusion is pure subjective probability (and liable to be a rotten approximation).

Evidently, many do not intend this meaning, and when they say "x is normally distributed" they imply that x is itself "alive" in some way, that there are forces "out there" that make, i.e. *cause*, x to take values according to a normal distribution.

Maybe the central limit theorem lurks and causes sums of individual grades, which form the GPA, to take certain values. This is incoherent. Each and every grade Sally received was caused, almost surely by a myriad of things, probably too many for us to track; and there is no indication that the same causes were at work for every grade. But suppose each grade was caused by one thing and the same thing. If we knew this cause, we would know the value of x; it would be deduced from our knowledge of the cause. And the same is true if each grade were caused by two known things; we could deduce x. But since each grade is almost surely the result of hundreds, maybe thousands—maybe more!—causes, we cannot deduce the GPA. The causes are unknown, but they are not *random* in any sense where randomness has causative powers.

What *can* we say in this case? Here is something we know:

$$\Pr(x = g_1|G) = \Pr(x = g_2|G), \tag{6.1}$$

where $x = g_1$ is shorthand for S = "Sally's GPA is g_1" (don't forget this!). This is the symmetry of individual constants, as seen in Chap. 4. G is equivalent to "We have a device which can take any of n states, $g_1, \ldots, g_n$, and which must take one state." From this we deduce

$$\Pr(x = g_i|G) = 1/n, \ i = 1, 2, \ldots, n. \tag{6.2}$$

There are no words about what caused any x; merely deduced information that the chance we see any value is as likely as any other value in the set of possible values. We could say that the uncertainty in x is quantified by a uniform distribution over $g_1, \ldots, g_n$, but since that leads to sin, it is better to say the former. Incidentally, a natural objection is that GPAs don't seem to be equally likely to be any number between 0 and 4, but that is because we mentally add to G evidence which is not provided explicitly. (I'm not claiming G is a good model.)

Can propositions have "true" distributions? Only in a limited sense. So-called random variables do not have to represent the "outcome" of the event from some experiment. Suppose X = "The color of the dragon is x"; if we let D = "Dragons can be green, black, or puce", the probability of "x" is easily computed, but we will never see the event. And there will be no real cause, either. This is the true probability, or true distribution, if you like. Any time we can deduce the "model", as it were, we have a true probability. But it is never the proposition that "has" a distribution or probability, it is only our understanding that does.

Lastly, when people think variables *have* "true" distributions, they are likely to blame data which does not conform to their expectations. Thus we see people tossing out "outliers". And since current practice revolves around model fit, data which does not fit increases the fit of what is left, leading to over-certainty.

6.5   Quantum Mechanics

A physicist designs an experiment according to quantum mechanical theory. This theory, or rather model, predicts that a certain quantum mechanical event, say a photon exhibiting circular or planar polarization, will occur with a certain probability. I discuss what models mean later, but for now this is clear enough. Given this model, we can deduce that if the experiment were run so many times, about so many circularly polarized photons will be observed. The experiment is run and it is discovered the model closely matches reality.

Experiments like this are of course conducted everywhere and often, with results closely matching reality being the norm. We therefore have strong confidence of the model's or theory's validity. But don't forget that in each and every case, without any exception whatsoever, some thing or things *caused* the outcomes of the experiment. If the photon is measured circularly polarized, something caused that to happen; if measured plane polarized, something caused that, too. What are these causes? Hold that question.

Given, "This is a two-sided coin which when flipped must show only one side" we deduce "The probability of H equals 1/2." We can run the experiment, as also has been done everywhere and often, and discover that, once again, our model matches reality. We have earlier seen that each flip was caused to fall as it did. But we usually don't know these causes. Because we do not know them does not mean they do not exist. Indeed, with coins not only do we know theses causes exist, but we can even, under certain controlled conditions, know the exact causes and thus deduce the events, i.e. predict them with certainty.

But not for quantum mechanical events. It appears we are barred from knowing causes beneath some level. Just like with the coin, though, because we do not know the causes does not imply they do not exist. Causes must exist. Things cannot happen for *no reason*, or *spontaneously*, which is sometimes a synonym.

Now there are claims that Bell's Theorem proves there are no causes of quantum mechanical events because Bell outlaws "hidden variables", which are taken to be the only possible kinds of causes for QM events. This is false, and must be false. Bell's arguments are probabilistic, and probability is an epistemological measure, not an ontological one, and so his proof, given his premises (assumptions), is about the state of our knowledge, or the lack of it. It cannot be that Bell has discovered places where the principle of causality is violated. Shimony has said [198] "*no physical theory which is realistic and also local in a specified sense can agree with all of the statistical implications of Quantum Mechanics.*" All this shows is that we cannot know what the causes are, or perhaps that the causes don't have a particular form implied by the theory; or it shows that the causes are "non-local". Yet on that point Murray Gell-Mann [90, pp. 171–172] laments that after experiments to test Bell were run "a wave of reports began to spread alleging that quantum mechanics had been shown to have weird and disturbing properties." What properties?

> The principle distortion disseminated. . .is the implication. . .that measuring the polarization, circular or plane, of one of the photons somehow effects the other photon. In fact, the

measurement does not cause any physical effect [i.e. cause] to propagate from one photon to another…If, on a particular branch of history the circular polarization of one of the photons may be measured, in which case the circular polarization of both photons is specified with certainty…no signal passes from one photon to the other in the experiment that confirms quantum mechanics.

Gell-Mann's intuition that causes are different than knowledge is correct. Consider this. Since everything that changes is at base quantum mechanical, or whatever it is that is "below" this (strings, say), then to say that every quantum mechanical events happens "spontaneously" is to say every single thing changes "spontaneously." That must mean *everything* that happens ultimately happens for "no reason." This is nuts. If there were no reason to anything, the world could no exhibit consistent structure. Science would be impossible, predictions would be of no use. The experiment ran above was carefully and diligently controlled, as quantum mechanical experiments tend to and must be. Single and paired photons are involved, and accurate predictions are made. But how to model complex entities like baseballs, trees, and human beings, how does everything fit together? That cannot be answered by relying on "randomness"

There are some authors, incidentally, like Fuchs who attempt subjective Bayesian probability QM theories. But since we have seen subjective probability is incorrect, I do not explore these further. This entire section, as is obvious, relies on the assumption that cause must be present, which is proved next chapter. I also do not attempt a complete explanation for QM here; the literature is vast and deep, and there obviously isn't the space here to survey it.

6.6 Simulations

These words from Jaynes are right: "It appears to be a quite general principle that, whenever there is a randomized way of doing something, then there is a nonrandomized way that delivers better performance but requires more thought."

We often hear of "simulating random normal" or creating "stochastic" or "synthetic" variables, or perhaps "drawing" from a normal or some other distribution. Such things form the backbone of many statistical methods, including bootstrapping, Gibbs sampling, Markov Chain Monte Carlo (MCMC), and several others. As with every other mistake about randomness, these methods are wrong in the sense that they encourage loose and even magical thinking about causality, and they are an inefficient use of time. If assiduously applied, reasonably accurate answers from these algorithms can be had, but they don't mean what people think and more efficient procedures are available, as Jaynes said.

The way simulations are said to work is that "random" or "stochastic" numbers are input into an algorithm and out pops answers to some mathematical question which is not analytic, which, that is, cannot be solved by pencil and paper (or could, but at too great a difficulty). Let's work with an example. One popular way of "generating normals" is to use what's called a Box-Muller transformation.

Any algorithm which needs "normals" can use this procedure. It starts by "generating" two "random independent uniform" numbers U_1 and U_2 and then calculating this creature:

$$Z = \sqrt{-2 \ln U_1} \cos(2\pi U_2), \tag{6.3}$$

where Z is now said to be "standard normally distributed." We don't need to worry about the math, except to notice that it is written as a causal, or rather determinative, proposition: "If U_1 is this and U_2 is that, Z is this *with certainty*." No uncertainty enters here; U_1 and U_2 determine Z.

As above, *random* or *stochastic* means unknown, and nothing more. Yet there is the unfortunate tendency to assume that "randomness" somehow blesses simulations. But since randomness means unknowingness, how can unknowingness influence anything if it isn't an ontological cause? It can't. It is felt that if the data being input to simulation algorithms aren't "random", or simulated to look "as if" they were random, then the results aren't legitimate. This is false. Since randomness is not a cause, we cannot "generate" "random" numbers in any sense. We can, of course, make up numbers which are unknown to some people. Example: I'm thinking of a number between 32 and 1400: to you, the number is random, but to me it is *generated*, i.e. *caused*, by my feverish brain.[1]

Since probability is a measure of information, computers cannot generate random numbers (nothing can). What happens, in the context of our math above, is that programmers have created algorithms which will *cause* numbers in the interval $(0, 1)$ (notice this does not include the end points); not in a regimented way so that we first see 0.01, then 0.02, etc., but caused with reference to some complex formula. These formulas which, if run long enough, will produce all the numbers between $(0, 1)$ at the resolution of the computer (some will be repetitions): infinite resolution is not possible.

Suppose this resolution is 0.01; that is, our resolution is to the nearest hundredth. Then all the numbers 0.01, 0.02, ..., 0.99 will eventually show up (again, many will be repeated; of course, we assume the programmer hasn't left a hole in this sequence). Because the numbers do not show up in sequence, many fool themselves into thinking the numbers are "random", and others, wanting to hold to the odd mysticism but understanding the math, call the numbers "pseudo random", an oxymoron.

If we want to use the Box-Muller algorithm, we can sidestep this self-induced (and unnecessary) complexity and simply write down all the numbers in the sequence, i.e. all the pairs in $(0, 1)^2$ (since we need U_1 and U_2) at whatever resolution we have; with our resolution, this is $(0.01, 0.01), (0.01, 0.02), \ldots, (0.99, 0.99)$ (this is a sequence of pairs of numbers, of length 9801). We then apply the determinative *mapping* of (U_1, U_2) to Z as given above, which produces (3.028866, 3.010924, ..., 1.414971e-01). What it looks like is shown in Figs. 6.1 and 6.2.

---

[1]The number, incidentally, is 32.32.

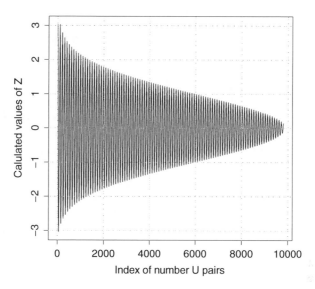

Fig. 6.1 The 9801 values of Z for each pair of (U_1, U_2) starting from $(0.01, 0.01)$ and progressing to $(0.99, 0.99)$

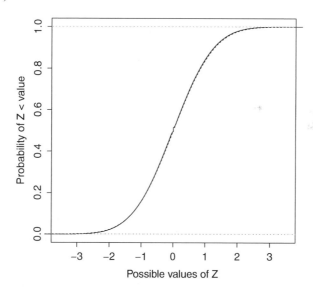

Fig. 6.2 The "true" value of the cumulative standard normal distribution in a dashed line, and the value of the ECDF for Z given by the mechanical approach

Figure 6.1 shows the *mappings* of the pairs (U_1, U_2) to Z, along the index of the number pairs. If you've understood the math above, the oscillation, size, and sign changes are obvious. In not, spend a few moments with this and it will become clear. Figure 6.2 shows the empirical cumulative distribution (ECDF) of the mapped

Z (solid), overlayed by the (approximate) analytic standard normal distribution dashed), i.e. the true distribution to high precision (as given by the R function pnorm which itself relies on an analytical approximation). It is difficult to see the deviation in the two plots.

There is tight overlap between the analytical approximation and the mechanical mapping, except for a slight bump or step in the ECDF at 0, owing to the crude discretization of (U_1, U_2). Computers can do better than the nearest hundredth, of course. Still, the error even at this rough resolution is small. I won't show it, but even a resolution 5 times cruder (nearest 0.05; number sequence length of 361) is more than good enough for most applications (a resolution of 0.1 is pushing it, but the reader should try it). This picture gives a straightforward, calculate-this-function, pen-and-paper-like analysis, with no strangeness about randomness—and it works.

What use is Z? Suppose we were after, say, "What is the probability that Z is less than -1?" All we have to do is ask: calculation from the ECDF involves counting the number of "generated" Z as less than -1 divided by (in this case) 9801, the length of the pairs of U. Simple as that. There are no epistemological difficulties with the interpretation, it comes right from the rule that all probability is conditional on the information supplied.

The analytic approximation to the probability $Z < -1$ is 0.159 (this is our comparator; calculated from R's standard approximation). With the resolution of 0.01, the direct method shows 0.160, which is close enough for most practical applications. A resolution of 0.05 gives 0.166, and 0.1 gives 0.172. I'm ignoring that we could have shifted U_1 or U_2 to different start points; I'm not attempting to provide optimal algorithms, only to show that the traditional "random" interpretation is wrong.

None of these answers have plus or minuses, though. With the 0.01 resolution, all we have is the answer with no idea of the size of the approximation error. In essence, there is no error, because the probability we have is conditional on the evidence we accepted. However, it is the evidence itself which is questioned. If this evidence is not logically equivalent to the function under consideration, then an approximation exists. The approximation is the accepted premises (resolution, the mapping, and so on) to the function of interest. Given our setup (starting points of 0.01 for U_1 and U_2, and the mapping function above), these are *the* answers. There is no probability attached to them, because none need be: we are certain of these answers given these premises. But we would like to have some idea of the error of the approximation. We're cheating here, in a way, because we know the right answer (to high degree), which in actual problems we won't. In order to get some notion how far off that 0.160 is we'd have to do more pen-and-paper work, engaging in what might be a fair amount of numerical analysis. Of course, for many standard problems, just like in MCMC approaches, this could be worked out in advance.

Contrast the determinative, fixed-pair method to the standard mystical or "simulation" approach. For the latter, we have to specify something like a resolution, which is the number of times we must "simulate" "normals", which we then collect and form the estimate of the probability of less than -1. This is done by counting like the fixed-pair method. To make it fair, pick 9801, which is the length of the 0.01-resolution series.

I ran this "simulation" once (using R's `runif` function and the Box-Muller transform) and got 0.162; a second time 0.164; a third showed 0.152. There's a new problem: each run of the "simulation" gives different answers. Which is the right one? They all are; a non-satisfying but true answer: they are all local or conditional truths. So what will happen is the "simulation" itself is iterated, say 5000 times, where each time we "simulate" 9801 "normals" and each time estimate the probability $Z < -1$, keeping track of all 9801 estimates? That kind of thing is the usual procedure. Turns out 90 % of the results are between 0.153 and 0.165, with a median and mean of 0.159, which equals the right answer (to the thousandth). It's then said there's a 90 % chance the answer we're after is between 0.153 and 0.165.

This or similarly constructed intervals are used as error bounds, which are "simulated" here, but could and should be calculated mechanically, as in the mapping approach. Notice that the uncertainty in the mystical approach *feels* greater, because the whole process is opaque and purposely vague. The numbers seem like they're coming from nowhere. The uncertainty is couched probabilistically, which is distracting.

It took 19 million calculations to get us the simulation answer above, incidentally, rather than the 9801 calculations (more or less) from the mechanical-causative approach. But if we increase the resolution to 0.005 in that approach, the answer is 0.159 at a cost of just under 40,000 calculations. Of course, MCMC fans will discover shortcuts and other optimizations to implement in their procedure; the 19 million may be substantially reducible. The number of calculations is a distraction here, anyway. Because we want to understand why the "simulation" approach works. It does (at some expense) give reasonable answers, as is well known. If we remove the mysticism about randomness and all that, we get Fig. 6.3.

The upper two plots are the results of the "simulation", while the bottom two are the mechanical-causal mapping. The bottom two show the empirical cumulative distribution of U_1 (U_2 is identical) and the subsequent ECDF of the mapped normal distribution, as before. The bump at 0 is there, but is small.

The top left ECDF shows all the "uniforms" spit out by R's `runif()` function. The only real difference between this and the ECDF of the mechanical approach is that the "simulation" is at a finer resolution (the first U happened to be 0.01031144, 6 orders of magnitude finer than the mechanical method's purposely crude 0.01; but the Us here are not truly plain-English uniform as they are in the mechanical approach). The subsequent ECDF of Z is also finer. The red lines are the approximate truth, as before.

Here's what's revealed by these pictures, the big secret: the "simulation" just *is* the mechanical approach done more often! After all, the same Box-Muller equation is used to map the "uniforms" to the "normals". That's the secret to the success of simulations: the two approaches, causal and simulation, are equivalent!

Which is now no surprise: *of course* they should be equivalent philosophically. We could have taken the (sorted) Us from the "simulation" as if they were the mechanical grid (U_1, U_2) we created and applied the mapping, or we could have pretended the Us from the "simulation" were "random" and then applied the mapping. Either way, same answer, which they had to be because there is nothing mysterious in "random" numbers.

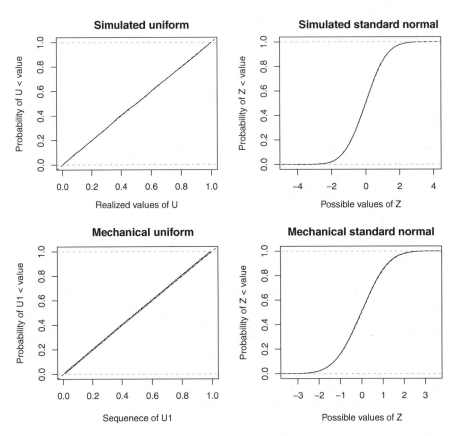

Fig. 6.3 The *upper-left plot* shows the realized values of U_1 from R's `runif` function (U_2 looks the same), and the *upper-right* shows the ECDF of the calculated values of Z (*solid*) and the true standard normal (*dashed*). The *bottom-left* shows the sequence of U_1 (U_2 is the same), and *bottom-right* shows the ECDF of the calculated and true values of Z. Both the simulation and mechanical approach give nearly identical answers to reasonable accuracy

The only difference (and advantage) seems to be in the built-in error guess from the "simulation", with its consequent fuzzy interpretation. But we could have a guess of error from the mechanical algorithm, too, either by numerical analysis means as mentioned, or even by computer approximation. One way is this: estimate quantities using a coarse, then fine, then finest grid and measure the rate of change of the estimates; with a little analysis thrown in, this makes a fine solution of the error. Much work can be done here.

The benefit of the mechanical approach is the demystification of the process. It focuses the mind on the math and reminds us that probability is nothing but a numerical measure of uncertainty, not a live thing which imbues "variables" with life and which by some sorcery gives meaning and authority to results.

6.7   Truly Random and Information Theory

One thing has to be admitted outright: that no field invents more tantalizing (and marketable) terms than computer science. *Machine learning, neural nets, universal approximators, genetic algorithms, artificial intelligence, fuzzy logic,* and the list goes ever onward. The promises implied in these phrases is inspiring. Machines that learn! Algorithms that figure out any problem with no human intervention! That a method fails to live up to its pledges is never remembered in the rush to embrace the next.

The empiricist bias of the methods is never noticed, either. Of course, computers work with "data", with empirical renderings of one kind or another, so that methods that are entirely empirically based are to be expected in practice. But we have already seen that some inductive reasoning extends beyond the empirical. Though many would argue the point, computers cannot do what rational minds can, for instance in induction-intellection. Recall in Chap. 3 that Groarke said induction-intellection provides the "Abstraction of necessary concepts, definitions, essences, necessary attributes, first principles, natural facts, moral principles." Though data to start these induction-intellections comes from empirical senses, it immediately (and instantaneously) extends beyond the empirical to the universal, to what can *never* be empirically verified. Computers, since they cannot think in this way, cannot do this.

Induction-intellection, induction-intuition, induction-argument, induction-analogy, i.e. the bulk of inductive reasoning, lie beyond the ability of any formal algorithm. Computer methods will thus never be a panacea for creating knowledge. Induction-probability is, however, is ripe for the picking, at least as far as the kinds of propositions in which we have an interest are observable, which we understand by now isn't always the case. The propositions of science are in large part empirical, so it is here we expect information theory and computer science to play the largest role.

One of the architects of information theory was Ray Solomonoff. His classic paper "A Formal Theory of Inductive Inference. Part I", [201], purports to be an existence proof for probabilities for "all problems in inductive inference", where he uses the term *induction* in its induction-probability sense, and where he does not appear aware that other senses of induction exist. Neither is the field to this date aware, as far as I can tell. In this paper (p. 16) he says his model—where he uses "model" in the sense of his scheme for computing probabilities and not in the sense used by statisticians—accounts for new observations in some sequence in an "optimum manner" (pp. 16–17):

> By "optimum manner" it is meant that the model we are discussing is at *least* as good as any other model of the universe in accounting for the sequence in question. Other models may devise mechanistic explanations of the sequence in terms of the known laws of science, or they may devise empirical mechanisms that optimally approximate the behavior and observations of the man within certain limits. Most of the models that we use to explain the universe around us are based upon laws and informal stochastic relations that are the result of induction using much data that we or others have observed. The induction methods used

in the present paper are meant to bypass the explicit formulation of scientific laws, and use the data of the past directly to make inductive inferences about specific future events.

It should be noted, then, that if the present model of the universe is to compete with other models of the universe that use scientific laws, then the sequence used in the present model must contain enough data of the sort that gave rise to the induction of these scientific laws. The laws of science that have been discovered can be viewed as summaries of large amounts of empirical data about the universe. In the present context, each such law can be transformed into a method of compactly coding the empirical data that gave rise to that law.

The hope is that an automatic method to discover all scientific "laws" (a term I discuss next chapter) is on the horizon. All we need is data of sufficient length, a computer powerful enough to hold it, and his algorithm which "automatically" applies probabilities, and all knowledge will be ours. But since "laws", such as they are, involve understanding causality, nature, and essences, and these acts of understanding are provided by inductions of forms other than induction-probability, this is a false hope.

What is to be applauded in Solomonoff's reasoning is his emphasis on prediction. He is not in the least interested in parameters, the obsession of statisticians, but only in what past observations have to say about future (rather, those not yet made known) data. We'll meet Solomonoff's probability formulation in the discussion of parameters in Chap. 8.

Related to Solomonoff's work is the idea of algorithmic complexity, and, with it, what information scientists call "random." Chief is the concept that we are working with a set of observed data, or a "string" of some fixed length written in some code (as the data of this sentence is written in English). We next take some model, or computer or real language, and express the string (observed data) in that language. The complexity of the string is the length of the shortest description of the string from the models under consideration. Loosely speaking, if this shortest description, conditional on the models, is no shorter than the length of the original data, the data is said to be "random". Another way: if the data can't be compressed by the model, the data is "random." Randomness is thus conditional on the model or models considered and is, as is clear, a synonym of unpredictable. Chaitin [43, p. 111] says, "There's only *one* definition of random…something is random if it is algorithmically incompressible or irreducible," and he humbly develops a measure of this which he calls "Chaitin randomness." Although his notation does not show it, Chaitin randomness, i.e. randomness, is conditional on the model or "machine" used, as notions of uncertainty always are. Again, unpredictable.

Knowledge provided by forms of induction like induction-intellection is thus random in this sense, since there is no way to get to this knowledge using any model. So axioms are random, too; all *sui generis* knowledge is random in this way. Randomness in the sense used by information theory is thus related to predictability. That which cannot be predicted with certainty, i.e. deduced, from some model, i.e. some set of accepted premises, is in some sense random; and this accords with the statistical meaning of the term. It is "beyond" the model or base of existing knowledge. The digits of π, for instance, are random in this sense, because their simplest description is just to list the digits. We know not from whence these digits come in any universal sense, given the premises that come in (for example) number theory. But the digits of π can be calculated to any finite expansion because algorithms exist. So π is not entirely random.

Finally, we can now see that so-called tests for randomness are misnamed. Since there is no such thing as randomness, tests for it are like tests for Bigfoot. Instead, what is tested for, and what should be acknowledged, is predictiveness. A sequence of numbers, or a string, or whatever, is more or less predictable. So what does *predictable* mean? That we have identified the premises which determine or which cause the sequence. Once this model is known, if it can be known, and we have seen in QM that not all models are knowable, but then again what causes axioms to be true is also not knowable, we can predict with certainty. Being able to predict without certainty is where uncertainty or "randomness" enters.

The amusing things about many tests for "randomness", i.e. predictability, is that they always turn a blind eye to the premises which are known to be determinative. One such algorithm is the Mersenne Twister. Its content is not of interest; what is, is that the sequence put out by it is, knowing the content and initial conditions, perfectly known. Tests for randomness are used on a given sequence, and these are said to be "random", but *only* because the content of the algorithm are ignored in the test! There are also firms that will supply, for a fee, "genuinely random" numbers, perhaps created through physical or mechanical processes. But since these don't exist, what is the customer getting? Simply a sequence which, examining only the sequence, does not allow certain predictions to be made of the (of future values of the) sequence. Of course, we have that sort of thing with QM. We can only predict within certain bounds, depending on the kind of experiment. And, again, the only limitation, but a big one, is that we are guaranteed not to know the causes behind the sequence. Since we can prove this by other means, there is no need to have a "test" for the randomness of such sequences.

Let me clarify that last, but utmost important, point. We often know what *determines* (i.e. ascertains) a necessary truth; these determinations are the basis of proof. But we can never know *why*, or rather, *what causes* these truths. Why are Peano's or any axioms true? Why a universe (where I use that word in its philosophical sense of *all* there is) like this, with these fundamental properties, whatever they turn out to be? I do not claim we know what we now call fundamental *is* fundamental in the same sense axioms are. I only ask why whatever is fundamental *is* fundamental. Answer: we have no idea, and we can have no idea. The mind of God is not ours to know. Necessary truths are the Way Things Are. And that is that.

Lastly, to clarify the clarification, here is an interesting point about "true" randomness that arose from a work by Donald Knuth [133]. Start with these equations:

$$e = \sum_{0}^{\infty} \frac{1}{n!}, \tag{6.4}$$

$$\pi = \sum_{k=0}^{\infty} \left[\frac{1}{16^k} \left(\frac{4}{8k+1} - \frac{2}{8k+4} - \frac{1}{8k+5} - \frac{1}{8k+6} \right) \right]. \tag{6.5}$$

The remarkable thing about (6.5) is that we can figure the n-th digit of π without having to compute any digit that came before. All it takes is time, just like in calculating the digits of e in (6.4). Now (the digits of) π and e are often said to be "random", e.g. [5, 12, 42, 155]. Since we have a formula, we cannot say that the digits of π are unknown or unpredictable. Yet there they all are: laid bare in a simple equation. I mean, it would be incorrect to say that the digits are "random" except in the sense that before we calculate them, we don't know them. They are perfectly predictable, though it will take infinite time to get to them all. But by "random" what is meant is that e and π are transcendental, meaning numbers that aren't algebraic, which in turns means that they cannot be explicitly and completely solved for. Yet Eqs. (6.5) and (6.4) solve for them in the certain sense that all the digits can be had if one is willing to wait long enough.

The equations here are determinative; they tell us the digits of e and π, and so these transcendentals are not random in a predictive sense since we have perfect predictability, but they are random in the sense that their origins are unknown. They don't tell us *why* it's these digits rather than some others. Nature is silent on the *cause* of these values. *Why* does $\pi = 3.1414593\ldots$? and not something else entirely? Answer: we do not know. It is the Way Things Are.

Chapter 7
Causality

"Anybody who writes a book in order to generate doubt on causality refutes the message by the very means that carries it."—Stanley Jaki.
"Felix, qui potuit rerum cognoscere causas"—Virgil.

A philosopher writes a book to convince his readers that causality is nonexistent. He hopes by his actions to cause his reader to adopt his view. How did the words get on the pages of the book demonstrating causality doesn't exist? The contradiction is never noted, perhaps because many expositors of theories exempt themselves from the consequences of their creations. Many modern philosophers are deeply suspicious about causality, a doubt reaching even to philosophical skepticism. Much of the distrust and misunderstanding of causality is because of post-Decartesian philosophy, which laid aside Aristotelian views prematurely, especially about the nature of cause. And this is odd because to Aristotle, the goal of science, and not necessarily its practice, is a matter of *rerum cognoscere causas*, or knowledge of the cause of things. This goal has largely been replaced by predictive ability in many of the sciences, which has much going for it, and which is a goal I suggest is returned to in fields which must use probability. But the ultimate aim of science must be the knowledge of the cause of things. If that is science, the predictive goodness and what is useful is not science *per se*, but techne or engineering.

That causality is doubted by some philosophers probably accounts for why many physical scientists routinely ignore philosophers. Scientists make their livings pursuing and even sometimes discovering (secondary) causes and can't countenance the idea that causes don't exist. Except for those scientists who are earnest in their attempts to say their measurements happened by magic, i.e. just happened for "no reason." On the other hand, some scientists enthusiastically believe all causal relationships can be discovered by applying the right computer algorithm or scientific "procedure". In an influential book, Judea Pearl writes [166], "The possibility of learning causal relationships from raw data has been on philosophers' dream lists since the time of Hume (1711–1776)." He, like Solomonoff in the last chapter, believes he has found this Statistician's Stone (this is my term). He hasn't

© Springer International Publishing Switzerland 2016
W. Briggs, *Uncertainty*, DOI 10.1007/978-3-319-39756-6_7

because it can't exist. And at any rate, learning causal relationships has been on philosophers' lists since the pre-Socratics: causality did not come into existence with Descartes, Hume, or Kant. And indeed, their view and the views of some other moderns is particularly stunted.

Cause is analogical. There is not one type or flavor or aspect of cause, but four: a formal, material, efficient, and final or teleological. Most causation concerns events which occur not separately, as in this before that, but simultaneously, where simultaneous events can be spread through time. Many causal data are embedded in time, and there two types of time series which are often confused: *per se* and accidental. These should not be mistaken for non-causal data series which are all accidental.

Causes, if they exist and are present, must always be operative, a proposition that has deep consequences for probability modeling. Falsifiability is rarely of interest, and almost never happens in practice. And under-determination, i.e. the possibility of causes other than those proposed, will always be with us.

Here is an example. Suppose scientists, via one of the NASA Rovers, found a device on Mars. It is roundish, the color of the Martian soil and occasionally displays, or is thought to display, what appear to be numbers. Scientists have decided the device has two "inputs", which are thought to be two protuberances in the "back". Through a series of inferences, it has been decided that the displayed numbers are correlated (I mean this word in its plain English sense) to the "inputs", which have been discovered to be "activated" (they flash different colors) in the same base of numbers as the display.

Put plainly, and I'll convert the numbers to base 10 for ease of understanding, the display is the sum of activations of input A and of input B. Mathematically, $A + B = D(isplay)$. So far, since the Rovers have not had much chance of observing the object, dubbed The Calculator, the activations of A and B have never been greater than 56 individually. I mean $A, B < 57$, which necessarily implies $D < 113$.

Naturally, since it is plain this is a device, scientists want to know its purpose. Theories are flying around NASA thick and fast. Though there are more theories than there are scientists, three rough camps have coalesced. Camp 1 says it's coincidental that so far $A, B < 57$, thus always $A + B = D$ no matter the number of activations of A and B. Camp 2 theorizes that the activations are "obviously" caused by two types of cosmic rays, which if they were to exceed some tolerance, they would cause $D = 5$. This, they say, is a derivation of string theory. I mean, if either $A, B > 56$, then the function is no longer a straight plus, but is instead a "quus"; i.e.

$$A + B = D \text{ , if } A, B < 57,$$

$$A + B = 5 \text{ , if } A, B > 56.$$

Camp 3 says $A + B = D$ for any number of activiations, but that after some period of time the device must start to degrade and that, because of various technical reasons,

$$A + B = D \text{ , before Date,}$$

$$A + B = C < D \text{ , after Date,}$$

and where the inequality is strict. We have plus and quus already, so call this (and this is my suggestion, not the scientists') "cuus".

The observations of the device are consistent with each of these three theories—and with many more theories, too. Recall I've only given you the three most popular. Obviously, none of the theories put forward by any of the scientists are inconsistent with the observations. We conclude from this that the physical observations are indeterminate; I mean, the state of the device, or the world plus the device, do not fully determine the device's purpose.

Still, even though the facts are indeterminate, we'd still like to know which of the plus, quus, cuus theories is right. I have no idea. Later (in Chap. 10) we'll learn that no theory "has" a probability, so there is no joy to be found in searching which of theories is more "likely". Of course, we could use each theory to make predictions and see which is better in some decisionable sense, which is very useful. But we'll never know which of plus, quus, cuus, or even some other theory, is true until we understand the purpose of the device. And we've seen that the facts alone do not and cannot determine what this purpose is.

To understand the purpose is to understand, in part, the cause of the device. Cause, as explained, is of four aspects: the formal or form, the material, the efficient, and the purpose, final, or end. In this case, the form is obvious enough: the device is "disguised" or made to look like a rock. The material is unknown at this point, but it's thought to be at least a rocky covering, or something which simulates rock. The efficient cause is, all agree, some kind of intelligence and whatever comprises the internal workings. Whether the designer is Martian or some clever human is unknown.

But what about the purpose of the device? Well, that's what the real unknown is. If it turns out that some Martian (or whomever) designed the device to count activations, however these are brought about, then plus is the right theory. If instead the final goal of the device was to count cosmic rays, then we're on to quus. Now it could be that quus and cuus are right, on the guess that the harsh cosmic rays are causing the degradation. That means the quus-purpose is right, but the cosmic rays efficiently cause a degradation which leads to cuus. So we have to be careful to keep in mind what part of the cause we're examining.

If somehow we discover the user's manual or tech specs for the device (and could translate them), then we'd know the cause of D—we'd know all aspects of the cause, and then we'd know the theory. And, as should now be obvious, what holds for this Martian device holds for all devices, whether made by Martians or via natural processes. It is only *after* we have knowledge of cause that under-determination ceases to be a problem.

Knowledge of cause is above, or rather beyond or deeper than, knowing what happens. Even beasts can know what happens, but they don't and can't understand why. Knowledge of cause is the grasping of essence, of the natures and substantial forms of the objects under consideration. None of these things are material in themselves, but are universals above and beyond the material world. Thus to come to knowledge of cause is to understand universals, which we get through a form of induction. Induction is the immaterial "movement" from finite particularities to an infinite generality and is such that only rational creatures can accomplish it.

The "quus" example is from Saul Kripke, as many will recognize. If not see [72]. Quus isn't usually presented with respect to under-determination, but of language and thought and how the intellect must be immaterial. I have concentrated on the epistemology, because uncertainty is our main interest.

7.1   What Is Cause Like?

In order to grasp *cause*, we need a brief, a very brief, introduction to the Aristotelian metaphysics of change. These are ancient views, once largely abandoned but becoming current one again for the very good reason they are correct. Philosophers like Nancy Cartwright [40], William Wallace [216], Ed Feser, and others are restoring a full and robust philosophy of Aristotelian causality back to the sciences. And there are other calls for scientists to sort out just what scientific pronouncements are: predictions or understanding of cause? See [41, 68]. What follows in this section is a précis of Feser's *Scholastic Metaphysics*, [74]. Full arguments are not given here, just enough information is provided to grasp the essential concepts; interested readers should follow up with the authors mentioned.

Contingent things, such as the book or "device" you are holding, exist as composites of act and potency, or actuality and potentiality. A lump of clay is *potentially* a vase. A lump of clay is not potentially a 1965 Barracuda with a 273 cu in. LA V8 (a weepingly beautiful automobile) nor is it potentially a stereo. A vase is *in potentia* to being a pile of shards. A vase is *in actuality* a vase, and a lump of clay is in actuality a lump of clay. The reader is *in potentia* to receiving a salary of fifty-thousand a year, unless he already possess that trait, and is therefore in actuality receiving it. And so on.

Some thing or things *must* cause *every* potentiality to become an actuality, that is, something actual must cause every *change*, where *every* change is an actualization of a potential. A potter is required to turn the potential vase *in* a lump of clay into a vase, while a child (in any of dozens of ways) can actualize the shards which are *in potentia in* that same vase, once it completed. Feser (p. 33): "These potentialities or potencies are real features...even if they are not actualities." Potentialities therefore exist in a certain sense, but *in potentia*. For instance, the number of numbers between 0 and 1 is *potentially* infinite, but not actually infinite in practice, a fact which has special consequences in measurement of real things.

Whatever is changed, is changed by another: whatever is in potential, is made actual *only* by something actual. Whatever cannot be changed, is not changed. It is not the lump's potential to be a vase that turns it in into a vase, it is an actual potter. The potter uses his *power* of making a vase; his hands are the efficient cause. The formal cause is the form of the vase, the material cause is the clay itself, and the final cause is the goal, the desire for the vase and not an ashtray. Clearly, the potter has the power to make the vase even when he is not making it (say, when he's taking his Barracuda out for a spin). Aquinas said,"nothing can be reduced from potentiality to actuality, except by something in a state of actuality" (*Summa Theologiae I.2.3*; quoted in Feser, p. 40). This is the *principle of causality* which I take as axiomatic. Things do not happen without causes, potentialities are not made actual by nothing, for nothing is not a thing, and that which is empty of everything has no power to cause anything. If things happened, i.e. change occurred, for no reason, then there would be no way to know that *this* change was a potentiality made actual by something actual or that *this* change happened for no reason or by magic. Batter steps up the plate and knocks on over the right field wall. Was this flight of the ball one of the times Nothing stepped in and did its non-cause trick which didn't cause the ball to take flight but which made it look like the ball was at one moment on the bat and the next soaring through the air for no reason? Or was this the time the batter gets the credit? Why bother doing science if you're not sure nature is going to cooperate or be whimsical? Change does not occur without it being caused. There is no magic.

Science deals with the contingent: (p. 106) a "contingent thing is such that its existence is distinct from its essence, where its essence is in potency relative to its existence, which actualizes it...To cause a contingent thing is thus to actualize a potency...*whatever is contingent has a cause*..." which is everything in science. This is not to say that *everything* has a cause; only that contingent things do, because only contingent things can be in potency. In *Summa Contra Gentiles* (Chapter 99, 2), Aquinas said, "Whatever sometimes is and sometimes is not, results from a cause: for nothing brings itself from not-being to being: since what is not yet, acts not." Only contingent things can sometimes be and sometimes not be.

A child throws a ball and it hits the vase. *As* the ball hits, the vase buckles; *as* the ball hits, the vase begins to break. The "event" is the ball-hitting-vase, and the ball hitting the vase event is *simultaneous*, which is not to say *instantaneous*. The ball hitting and the vase buckling happen over a short period of time; they are not different events "entirely loose and separate", to use Hume's mistaken phrase: there is *one* event, the simultaneity. It is not because we "happen" to see, or "chance" upon the spectacle of ball-hitting-vase that we know the ball *caused* the vase to break. It is because we learn, via induction, that balls traveling at sufficient speed have the *power* to break vases of this certain type. It is the vase's nature to break when hit by balls like that under these circumstances. We are back to essence. Understanding essence and powers is to understand cause.

Many modern authors put this the wrong way, saying *first* the ball hits *then* the vase breaks. This is not so. There are not two separate events, but one joint event, spread through time. This point is crucial. It is difficult to find modern examples where distinctness in events and separateness in time is not assumed. Of course, that the ball-hitting-vase is spread through time, however brief, does not mean that all events are. Certain quantum mechanical events are thought to be *instantaneous* (but proof of this is lacking; *instantaneous* is a remarkably strong attribute). But that merely confirms the view that we are not witnessing "loose and separate" events, but joint ones.

Knowing the ball was the efficient cause of the vase breaking is not the whole story, though it is enough for most (it was for my mother). There are all sorts of forces involved, including the ball's momentum, friction, elasticity of both objects, and so forth. These are *not* necessary to understand to say the ball caused the break. These additional forces can be investigated to form a deeper understanding the precise mechanisms and powers and to, say, knowing when the vase will or won't break. Each of these micro-investigations, as it were, are no different than the gross version. The essence and powers of the forces involved are understood to be causes. But there are limits to our knowledge.

Let's investigate the ball-hits-vase joint event more closely. The ball and vase are not monoliths, but composed of smaller parts. *As* the ball pushes into the vase, the molecules of the ball and vase are themselves undergoing change. These changes, which are actualizations of potentials, are caused by something actual, which are the atoms in the molecules. These are also undergoing change, which are again actualizations of potentials, which are also caused by something actual. This might be the interactions of the constituents of the atoms, the electrons, protons, and neutrons, which are also undergoing change. That means there are more actualizations of potentials caused by other somethings which are actuals. These may be quarks, which are themselves pushed about by (say) actual strings (or super-strings), which themselves, perhaps, are caused to change by something "below" (at a more fundamental level than) them. All of this is happening *here-and-now*, simultaneously, but again not necessarily instantaneously. All of these actualizations of potentialities by other actualities is called a *per se* times series, or a *per se* series of events in the here-and-now time.

But you can see that this process cannot continue to infinity. It must bottom out, or nothing can ever get moving; no changes could ever be made. There must be some first cause or first mover or first changer. This makes all other causes in the chain *secondary causes*. Secondary causes are the subject of physics; the first or base cause belongs to metaphysics. The first cause must be entirely actual and have no potential. It is what makes all "bottom" potentialities actual. It is responsible for every contingent event, at base. This is the *prime* or *primary* cause, which is ever-present. Science is and must forever be ignorant of this cause; that is, of the *why* of this cause, or how this cause is decided or acts. A *per se* series is a handy explanation of quantum mechanical EPR-like events, or whatever is "beneath" them, as discussed earlier. Again, all of the other here-and-now causes—string into quark into protons into etc.—are *secondary* causes. All have powers and essences, and it is the goal of science to understand these secondary causes.

There is another type of causal series, this one distinct in time, an *accidental* series. The classic, and really perfect, example is that a grandfather caused his son to be made and he, your father, caused you to be made. This doesn't stop with your grandfather, naturally, but continues along a string of relatives into the past (and perhaps into the future, if you are so blessed). Remove one of the knots in the string, i.e. remove one of the causes, and you would not be reading this now.

There are also non-causal accidental series. Unfortunately, in practice, data analysts often think of accidental series as if they were causal. The field of time series analysis comes to mind. Examples of non-causal accidental series: yearly (or monthly or daily or hourly or whatever) average temperature (or sales figures or unemployment rates or suicides, or etc., etc.). Last year's average did not and could not *cause* this year's average. How can an average, a mere weightless number, cause anything? Yet these kinds of series are often supposed to be causal. Result? Misascribed causes and wild over-certainty. I leave discussion of these accidents until the last chapter.

7.2 Causal and Deterministic Models

A causal model is a collection of premises from which are deduced a set of propositions that are certainly true or false and where we have an understanding of the powers and essences of the objects considered. Without understanding of the powers and essences, the model can be no better than deterministic. For example, given "At time t the object will be red, else blue" the proposition "It is not time t and the object is red" is false. The object may be red at times other than t, and, if so, the model is falsified. But this is not a causal model, because we don't have any understanding of the nature or essence of the color change. The premises are enough, however, to *determine* the change. We must be ever on guard of the analogical nature of the word *determine*. To make the model causal, we'd have to add the "why" of the color change.

The model may, of course, be perfectly predictive but that does not make the model causal. Why? Notice carefully that there is nothing in the model that describes the efficient cause of the object changing colors.

There are four types or kinds of causes: formal, material, efficient, and final. We can make the model causal by grasping the nature of the object and powers of the thing bringing about the (secondary) cause. The object of which we speak must have a certain form. Say, a red lollipop. The form of this sucker, as we called them in Detroit, is constructed of some material, usually sugar, chemical coloring and flavoring. Some thing took these materials and put them into the form we see; this might be an assembly line with its associated machinery. This machinery was the efficient cause. In the end, we eat the thing, which is its purpose or end; rather, the end was the creation of an edible piece of candy.

In the first example (a very weak model) there are no premises about the object changing form other than its color: it may change shape as well, but only to the extent that we still recognize the object as *the same* object. It's not clear whether the color change will be the addition of new material (say, ink), the subtraction of another (perhaps by sun bleaching), or because it was the nature of the object to be red at times or blue at times, a change triggered by who knows what efficient cause.

A simple model is "If X then Y". When somebody asks "Why Y"? The indubitable answer is "Because X." This model is as simplistic as can be, but it is not trivial or empty. All our knowledge provided inductively, like axioms, are given by this form, where, of course, X might be a compound statement. On the other hand, these are not truly causal models: these are not even explanations. *Why* is or *how* is it that the (say) principle of non-contradiction is true? We cannot say. *It just is.* We can say that we *know* it is true given our scant observations via induction. But that is an epistemological explanation and not a causal explanation. We cannot know why or how something that is necessarily true is necessarily true. That kind of understanding, like knowing the full nature of the first cause, is closed off to us. There are some things we must accept on faith.

True causal models instead relate to secondary causes, of the type mentioned above. Y = "The vase is in shards." Why Y? Be*cause* X = "The ball hit it," where

X is shorthand for the forces we know to be responsible. This, too, is a weak model because it only applied to this ball and this vase. But we can broaden it to all balls and all vases under specified conditions. This can be informal, as nearly all of our causal models are, or formalized with mathematics. The danger with mathematics, as ever, is the Deadly Sin of Reification, when we give life to the equations and forget they represent real objects. Also, we cannot mathematize all parts of most real-life events, we can only create abstractions from them. Reification happens, all too often, when we forget that our creations are abstractions and not reality.

A fuller example. The equation for the height y of a projectile is given as $y = \tan(\theta) \cdot x - g(2v_0^2 \cos^2 \theta)^{-1} \cdot x^2$, where θ is the initial angle of the projectile launched with initial velocity v_0, gravitational acceleration g, and the distance from the (arbitrary) origin of the throw x. This is a deterministic model. It says that *given* g, v_0, and x, y will be such-and-such a value with certainty. But it doesn't say what *causes* y; it only says the value of y is *determined*, i.e. *made known*, because of these other things. To understand the efficient cause of y, we must go deeper. The nature of the "projectile" is vague enough, but we understand that it is the nature or power of the impelling force, and of the other forces, to cause the projectile to scuttle along. It could be a coincidence that this equation, which yields perfectly reasonable, or even completely accurate, predictions, is itself the result of other causes that are the real cause of y. We only know to stop this kind of thinking after induction tells us we have understood the essence or nature of the situation.

The premises of this model are explicit and not subject to "fuzzy" interpretation. One premise is "This precise single fixed g", not "This g more or less." The same is true of the proposition itself: y will be this and only this value, not this value plus-or-minus or more-or-less. Add "close enough" conditions to the premises and the model retains its deterministic status; adding them to the proposition of interest turns the model from deterministic to (perhaps partly) probabilistic. Many deterministic models are treated probabilistically when the propositions of interest are observed to be false. If, given the premises, the projectile is not precisely at y, but at y more-or-less, it is usually thought that the main model is doing "most of the causing" and therefore there must exist unobserved or unnoticed, or at least unconsidered, causes that also operate on the projectile which are not in the model. This is fair enough when the nature or essences of the main causes are certain, but we really have a probabilistic model.

Finally, if we do understand the cause of some thing, we don't need a model or experiment. Why? Because we know the cause! We only need deterministic models (like the projectile equation) to understand the *extent* of a cause in a situation which specified conditions. This is, of course, a trivial observation, but it will have an important sequel.

7.3 Paths

There is a difference, as there was for truth (necessary and conditional or local), between universal and partial or limited deterministic models. The model of the projectile was, in absence of any other information, partial; so was the red-blue-object model. Both models say propositions will be true or false given the stated conditions, but the partial model contains premises which are not (known to be) necessarily true. Deterministic models may also be over-loaded, which is when two partial deterministic models have different, not logically equivalent premises, but which make identical predictions about a set of propositions. More than one model can explain the same set of facts. But there can only be one true understanding of cause.

The goal is to discover universal deterministic models, which contain necessarily true premises and which lead to certainty and where the nature and essence of the events are understood. Given the results of quantum mechanics, it appears this goal cannot be met for efficient causes for some events. No full, universal efficient deterministic model of nature exists: if one did, it would be the prized Theory of Everything. Even though we are barred from complete knowledge, rich and useful conditional models abound.

Einstein, Podolsky, and Rosen [67] famously said that "If, without in any way disturbing a system, we can predict with certainty (i.e., with probability equal to unity) the value of the physical quantity, then there exists an element of reality corresponding to that quantity." What might that mean to deterministic versus causal models?

Before you is a machine that has a dial marked 1 through 3 and a light. Moving the dial through its states and the light turns yellow, blue, white. From this you form the premise (with obvious shorthand) "If D1, yellow; if D2, blue; if D3, white." This is a deterministic model. It says that, given certain conditions, certain other things happen with certainty. Extreme probabilities (0 and 1) are easily derived from deterministic models with the addition of a minor premise and some proposition of interest. For instance, add the minor premise "D2 (the dial is in position 2)" and propositions "The light is white" or "The light is chartreuse." Given this model, these propositions are false. We could have also deduced, from these two premises, the proposition "The light is blue."

Why did the model turn its various colors? I have no idea. How can the model be causal if we don't know all the causes of some event? Because it turns out we don't know all (as in *all*) the causes in any contingent event, yet we can sometimes understand essences. I don't need to know, or even need to care about every cause of the light, either, not if all I am interested in is its color.

The model relates to propositions of the light's colors, even though there are lots of facts about the machine and it milieu which exist but which we ignore. It is I hope obvious that some thing or things were the efficient cause of the light turning color. The dial played a role, but given our understanding of physics, we suspect it wasn't the dial itself that made the light glow. The light glowed because certain elements were electrified as opposed to others, and the electrification was caused by the states of certain resistors, diodes, etc. And the states of those resistors etc. in turn were caused to be in those states by their chemical components being in a certain way. And so on down the chain, all the way to the bottom (which we saw above must exist).

That is, the best we can do is to end at uncertainty. We don't know how the Ultimate Cause works; we don't even know how electrons work, not entirely, and seem to be forbidden that knowledge. Quantum mechanics is, of course, a theory of uncertainty. And since uncertainty is a measure of knowledge, quantum mechanics is not a causal theory, but it is a probabilistic-deterministic one. We don't know why this particle takes this spin rather than that spin. We *can* say, given a set of premises, the probability it takes this spin. But this is not to abandon causality. *Some thing* or things must be *causing* whatever happens to happen, something must be reducing the potentiality of being in the state spin up to being actually in that state. Quantum mechanics events cannot happen for *no* reason. It cannot be that *nothing* causes these events. How could it? It's *nothing*. Nothing is the *complete* absence of *anything*. To say that events happen "spontaneously" in the sense of "from nothing" or "uncaused" is to fool oneself and to embrace a kind of mysticism or magic. EPR's "element of reality" can be interpreted as the actuality, whatever it is, that actualizes the potential. Bell's Theorem (appears to) prove that we can't know what this is, but it cannot prove that it doesn't exist. Nothing has no power.

Why this is the case was explored previously. The point here is that if we knew all the conditions and causes of why the light turned yellow, we would have a full or universal causal model. Since instead we have only limited knowledge (the dial positions), we have a local deterministic model. The terms *universal* and *local* apply to the completeness of the model. Since we never know everything, all scientific models are thus local. No matter how well we understand any system, we will never be able to understand why what is happening at the most foundational levels is happening. So whenever we speak of causal models, we're always leaving something out. The degree to which we leave things out is great or small, but there is no difficulty is saying the dials "caused" the light to turn various colors, or that the "x" caused the "y" in the trajectory equation, as long as we keep in mind this is shorthand.

A local deterministic model is a collection of premises for which any proposition in relation to this collection is certainly true or false or irrelevant. Since all we ever have in science are local models, henceforth I'll drop the "local." The irrelevancy is necessary. Suppose we have our dial model and desire the (conditional) probability of "Mike likes tacos". How do we know this proposition is irrelevant with respect to this model? Why aren't we conducting an experiment to verify this? Because we discern the essence of the machine and of people like Mike, and we know via induction that the two things have nothing to do with one another. We can never escape induction.

But this isn't quite the right flavor. Actually, in models, we usually start out with the propositions of interest and search for premises which make these propositions true or likely and others false or unlikely. The propositions of interest with our machine is "The light is yellow", etc. Models which make any proposition true are trivially found. For instance, given "The light is yellow" it is true that "The light is yellow." But we don't accept this premise in our model because we are after understanding of causes. Adding the trivial premise does nothing to further our knowledge.

Accept the same model: now what if the light glowed red? Then our model is *falsified* because (our information insists) the dial must be in one of the three positions, and these we have said certainly lead to other colors. Suppose the dial in fact was in position 2 when the light turned red. We can then modify our model in the obvious way, allowing for the possibility of two colors in position 2. This turns the deterministic model into a probabilistic one. It also makes a *new* model. This point cannot be too highly stressed. Any change to a model such that the probabilities or certainties of the outcomes (propositions we put to the model) also change creates a *new* model, even though it might, through custom, retain its old name. We cannot say in general that adding new premises creates a new model, because we can always add necessarily true premises or even possibly irrelevant ones without changing the probabilities of propositions.

We needn't capture everything that can happen to have a causal or deterministic model. For instance, here is an alternate model "If D1, yellow, otherwise something else." This model works. It is even as accurate as the first model! Whether it is more or less useful, however, depends on decisions and actions taken based on the model. It is obviously less precise, but less precision can be a blessing. We might have thermometers which measure air temperature to the nearest hundredth degree, but most people will be content to knowing the value within two or three degrees (Fahrenheit). Indeed, in all problems, the measurements and models should be no finer than the decisions to be made. This point is taken up again later.

One reason why it is thought probability models can discern cause, especially in the computer sciences, is a because of an unrecognized bias. As of this writing, a hot topic is "deep learning", which can be described as machine "learning" iterated, a suite of computational tricks to find "signals" buried in "noise." The hope is that as datasets increase in size and complexity automated (deep "learning" or otherwise) algorithms will discover the causes of "outcomes"; see e.g. [158]. The bias is exposed by thinking about who decides what goes into the databases as potential causes or proxies of causes. Consider the proposition "Bob spent $1,124.52 on his credit card." This "effect" might have been caused by the sock colors of the residents of Perth, say, or the number of sucker sticks longer than three inches in the town of Gaylord, Michigan, or anything. These odd possibilities are not in databases of credit card charges, because database creators cannot imagine how these oddities are in any way causative of the effect of interest. Items which "make the cut" are there because creators *can* imagine how these items are causes, or how they might facilitate or block other causes, and this is because the natures or essences of these items are known to some extent. The form of the blocking (see below) or the conditions of when the cause operates might not be known, but that an item plays some role in the causal process is at least suspected.

7.4   Once a Cause, Always a Cause

This section is relevant for all statistical and probability models which form the conceit that they have identified the cause of some data; the material is based on [28]. Suppose we learned that 1000 people were "exposed" to PM2.5—which is to say, particulate matter 2.5 microns or smaller—at some zero or trace level, and that another group of the same size was exposed to high amounts. Call these two groups "low" and "high PM2.5". Suppose, too, it turns out five people in the low group developed cancer of the albondigas, and that 15 folks in the high group contracted the same dread disease. (If you don't love this example, substitute placebo versus drug or some other on-and-off, yes-or-no dichotomous state.)

What *caused* the observed difference in cancer rates? Some thing or things caused each unfortunate person in our experiment to develop cancer. What could this cause or these causes be? Notice I emphasize that there may be more than one cause present. It needn't be the same thing operating on each individual. Each of the 20 people may have had a different cause of their cancer; or each of the 20 may have had the same cause. And this is so even though it may be that cancer of the albondigas is caused in the human body in only one way. Suppose some particular bit of DNA needs to "break" for the cancer to develop, and that this DNA can only break because of the presence of some compound in just those individuals with a certain genetic structure. Then the cause or causes of the presence of this compound become our main question: how did it come to be in each of these people? That cause may be the same or different.

There is no proof in the data that high levels of PM2.5 cause cancer of the albondigas. If high levels did cause cancer, then why didn't every one of the 1000 folks in the high group develop it? If high PM2.5 really is a cause—and recall we're supposing *every* individual in the high group had the same exposure—then it should have made each person sick. Unless it was *prevented* from doing so by some other thing or things; e.g. perhaps a counter-balancing cause operates that acts "oppositely" of PM2.5. High PM2.5 cannot be a complete cause: it may be necessary, but it cannot be sufficient. And it needn't be a cause at all. The data we have is perfectly consistent with some other thing or things, unmeasured by us, causing every case of cancer. And this is so even if all 1000 individuals in the high group had cancer.

This always-or-nothing is true for *every* hypothesis; that is, every set of data. The proposed mechanism is either *always* an efficient cause, though it sometimes may be blocked or missing some "key" (other secondary causes or catalysts) or be counterposed by some other cause, or it is *never* a cause. There is no in-between. Always-or-never a cause is tautological, meaning there is *no* information added to the problem by saying the proposed mechanism *might* be a cause. From that we deduce a proposed cause, *absent* knowledge of essence, said or believed to be a cause based on some function of the data, is always a prejudice, conceit, or guess. Because our knowledge that the proposed cause only might be always (albeit possibly sometimes blocked) or never an efficient cause, and this is tautological, we cannot find a probability the proposed cause *is* a cause—conditioned only on that tautology, that is.

Consider also that the cause of the cancer could not have been high PM2.5 in the low group, because, of course, the five people there who developed cancer were not exposed to high PM2.5 as a possible cause. Therefore, their cause or causes *must* have been different *if* high PM2.5 is a cause. And even if PM2.5 is *a* cause, it is not necessary the only cause. The same cause that operated in the low group, or some other cause entirely, might have struck some or all of the afflicted in the high group. In other words, since we don't know if high PM2.5 is a cause, we cannot know whether whatever caused the cancers in the low group didn't also cause the cancers in the high group. Recall that there may have been as many as 20 different causes. We conclude that nothing in the plain observations is of any help in deciding what is or isn't a cause. That statement has tremendous importance when considering standard statistical procedures.

Given the multitude of possible measures we can make on actual people—everything from whatever they've eaten over the course of their life to the environments to which they have been exposed, and on and on almost (but never in reality) endlessly—it is more than reasonable to suppose that we can discover some thing which is also different between the two groups besides exposure levels. Suppose it turns out—and something like this almost surely will—every person in the high group ate at least one more banana than did folks in the low group. That means whatever conclusions we reach via some statistical analysis, we could have equally well put down to having eaten more bananas. This is because the label "low PM2.5" and "high PM2.5" can be swapped for "low banana" and "high banana", a set of measurements just as true and valid. Call this the banana test.

Clearly, there was *some* thing or *some* things different between the two groups. There must have been, because the number of people who got cancer was different, and the difference was caused, as must be true. But there is absolutely nothing in the observations alone that tell us what this cause was or what these causes were. We are not just discussing PM2.5. The criticisms here apply to *every* classical statistical analysis ever done.

Yet there is plausible suspicion that PM2.5 and not bananas *might* cause disease. We know this because we suspect it is in the nature of fine particulate matter to interact with, and possibly interfere with, the functioning of the lungs, the nature of which we also have some grasp. We do not *know* just based on the raw data—and never forgot that we can only know what is true: though we can *believe* anything—that PM2.5 causes cancer. A reasonable condition, given what we have learned from other dose-response relationships, is that greater exposure to PM2.5 will give more opportunity for whatever it is in PM2.5 that causes cancer to operate. But we don't have that in this experiment. So we can only *assume* PM2.5 is a cause and make verifiable predictions to test this assumption.

Notice that in this approach we *must* assume that (high) PM2.5 is *always* a cause but that sometimes it is stopped from operating because of some lack: say, a person has to have a specific genetic code, or must inhale the dust only when breathing is labored, or some chemical must be present, or whatever—the exact conditions may be exceedingly complex. As we saw above, the only other assumption is that PM2.5 is *not* a cause, and if it is not, then we *must not* use a probability model supposing PM2.5 is a cause.

This implies the following curious result. Probability models aren't what you might have thought. If we assume PM2.5 is a cause, then we must conclude that it is sometimes blocked, else all 1000 in the high group would have become ill. And recall that if we assume PM2.5 is a cause, it *necessarily* implies there is at least one other cause, a cause which must exist to account for the illnesses in the low group. Saying PM2.5 is a cause thus creates a mystery: what is this other cause (or causes)? But it also means that the probability model in the high group is *not* a model of cause: it is a model of *blocking*. The probability models doesn't say, not really, "This person has a this-or-that chance of developing illness if exposed to PM2.5", rather, "The chance the causal effect of PM2.5 is *blocked* is this-and-such." And even that pronouncement is still conditional on believing the other cause or causes besides PM2.5 don't operate in the presence of PM2.5, and where is the evidence for that? There is none. Probability models always belie uncertainty. They are never proof of cause, which is why automated attempts to "prove" cause in large collections of data, e.g. in [166], must fail. Uncertainty always lingers unless there is knowledge of power and essence. Probability models themselves are explored in depth next chapter.

7.5 Falsifiability

There is passed around a corruption of a quip by George Box that runs "All models are false" or "All models are wrong." What Box actually said was "Remember that all models are wrong; the practical question is how wrong do they have to be to not be useful." This is an instance where shortening helps the grammar. Either way, the sentiment is false: all models are not wrong.

All models cannot be false. Indeed, something like the opposite is true: given their premises, all models are valid, assuming no errors in calculation or application, of course. Not all models are sound, but many are. Not all models are useful, but some are. A simple model is a coin flip. Given "This is a two-sided coin with just one side labeled H, which will be flipped and which must land on only one side" the probability "The coin lands H" is 1/2. There is no word in this model about what *causes* the coin to land H or T. These causes must always exist. There is also no way to deduce from the premises that this model "fits" or "works" with real coins. Change "coin" to "interocitor" to see this. The probability remains the same—the model is valid—but there are no interocitors.

Falsified has a precise, unambiguous, logical, mathematically rigorous meaning: that something was shown to be *certainly false*. Given $x + 2 = 7$, it is false certainly that $x = 2$. There is no ambiguity here. If a deterministic, causal, or probability model says that Y is (or will be) true, i.e. it has 100 % certainty, and it turns out upon observation that Y is not observed, then the model is falsified. If any model said, "The probability $Y = y$ is 1" and Y is observed to be anything but y, the model is falsified. But if the model said "The probability $Y = y$ is ϵ" and Y is observed to be y, the model is *not* falsified. If anything, the model is *verified*, since the model said Y could be y, and it was observed to be y.

According to Karl Popper, and widely believed by many, a theory is said not to be scientific unless it is falsifiable. This is an understandable definition, but as something philosophically useful it fails because most theories scientists hold are not falsifiable because of their, at least semi, probabilistic nature. The call for falsifiability retains its appeal, however, because those who champion flawed theories are annoying and we desire a weapon to dispatch dumb theories.

A theory or model is a set of propositions which are taken or assumed true. Presumably these propositions are not self-contradictory, though in complex theories, who can say? It is not necessary that any one person know each of the propositions, or even that the set is closed (as we'll see). A complex theory contains more propositions than a simple one. There is no need to be more precise than this except to stress that a theory or model is its premises. Changing any one of them changes the theory into a new theory.

We hear things like, "Given my theory of the weather, tomorrow's high will be 70°F." This is usually shortened to "Tomorrow's high will be 70°F" which the conditioning left implicit. Now if the high temperature tomorrow is anything but 70 °F, the theory is falsified. The theory said some thing would occur with certainty: it did not: the theory is false. End of story.

But nobody understands the phrase "will be 70°F" to mean "will be precisely, exactly, to the nth decimal place 70°F." Words mean something. Our task is to translate the prediction into the vernacular. That means adding some "fuzz" around 70 °F; or, in other words, by taking the phrase to mean "There is a good (but not perfect) chance the high will be 70°F". And, of course, "good (but not perfect) chance" puts us on probabilistic grounds.

In this case, the theory only said something might happen. If it didn't, the theory is not falsified. How can it be? One of the things the theory said *could* happen *didn't* happen. This cuts both ways. If 70 °F obtains, the theory is not completely validated, either. That is because, adding uncertainty, the theory might have also said it could have been 69 or 71 °F, albeit with a smaller chance, and these did not occur. Whether the theory is useful depends on the decisions we make given the prediction; which is to say, on how we have (or would have) acted on the predictions as they stand (or stood). This is an entirely different topic, the gist of which is that a theory useful to one man may be useless to another. This topic is explored later.

You have it by now: if the predictions derived from a theory are probabilistic then the theory can never or rarely be falsified. This is so even if the predictions have very, very small probabilities. If the prediction (given the theory) is that X will only happen with probability $\epsilon > 0$, and X happens, then the theory is not falsified. Period.

Most theories, even though stated in deterministic or causal terms, are actually meant, and are surely taken, in a probabilistic cast like the temperature forecast. This is because of the presence or suspicion of measurement error, imperfect specifications, and a host of other reasons which accompany any theory in practice. Entire fields are nowhere near falsifiable in the sense that the predictions associated with them are probabilistic or "fuzzy": biology (the theory of natural selection and every single evolutionary psychology theory), sociology, economics, psychology, education, and such forth. But even a broad range of theories within more rigorous fields, like physics and chemistry, are also not falsifiable in the practical sense. (The reader can apply on his own the arguments given here to discern whether his favorite theory can be falsified.)

Again, the sole way a theory can be falsified is if it states, in no uncertain terms, boldly and forthrightly, adamantly and insistently and uncompromisingly that X cannot happen, that the probability of X is 0, exactly 0. Then if X happens, *à la mort*, else not. Falsified is akin to mathematical proof: it is undeniable.

All theories of the contingent are therefore trivially falsifiable (in logic) for some propositions. For example, we can derive from most any theory the prediction that the probability X = "The existence of a twelve-and-a-half footed half-duck half-snake that speaks (what else?) French on Uranus" is 0. But since we have to travel to that frigid locale in order to verify X, we will never learn whether the theory is falsified in fact because, of course, since X is contingent it might (conditioned merely on the premises which identify its contingency) be true. Therefore, we are interested in falsifiability in practice.

One last thing about probability models and falsification. If a probability model, conditional on whatever evidence it has, says X will happen with some non-zero probability and X is never observed, then the model is *not* falsified. Now many statistical models employ the so-called normal distribution, such as in regression. Later I'll use a grade point average example. GPA cannot be less than 0, but a normal-regression will *always* given probabilities to values less than 0 (incidentally, this is called probability leakage, [26]); indeed, this model gives positive probability for any conceivable interval. Thus no observation can ever falsify it. If we want to say how good this model is, then, we'll have to look for another way. This is provided using the concepts of verification, scoring, and skill, given in Chap. 9. Given that many physical models often have subjective fuzz to them when they fail, it is no wonder that falsifiability has never proven to be a useful concept in practice.

7.6 Explanation

Inference to the best explanation, a.k.a. abduction, is often defined with respect to surprise. But all surprise is conditional on tacit or already accepted knowledge. Your dad comes into the house, his clothes smeared with red paint. What is the explanation for the paint (and not your dad entering the house *per se*; as always, we pick the propositions of interest)? Well, it could be that he was out painting the house. Or perhaps he was painting the car, or maybe even an old chair in the garage. Or it could be that scamps drove by your house and paint-bombed him. Or perhaps he was abducted by Martians and the splotches are their probe markings. And so on, endlessly.

Pick one of these explanations and suppose it's the only one we have. Say E = "Dad was out painting the house". The observed data is D = "Paint-smeared clothes." There must be *tacit* premises floating about, something to connect painting with paint on clothes, like "One who paints often but not always gets smeared," or a plain "One who paints [always] gets smeared". Whichever, E is the best explanation. Why? We have inferred the best explanation because it was the *only* explanation. Probabilities like Pr(D|E) and Pr(E|D) are irrelevant, and incorrect. This point is often confused because of miswriting Bayes's theorem. Some incorrectly, or rather incompletely, write

$$Pr(E|D) = \frac{Pr(D|E)\,Pr(E)}{Pr(D)}. \tag{7.1}$$

The mistakes are "Pr(E)" and "Pr(D)", which do not exist (they have no conditions). It might be tempting to argue the denominator can be expanded:

$$Pr(D) = \sum_i Pr(D|E_i)\,Pr(E_i),$$

but this repeats the error (for every i) in writing $Pr(E_i)$, none of which exist. No unconditional probability exists, as proved earlier. The same mistakes would be made were D and E swapped in these equations.

There is no way to say how surprising D was *except* by reference to some explanation. To think otherwise is commit the fallacy of p-values in a Bayesian context (e.g. [20, 62, 91, 157]). Suppose that Pr(D|E) is small but non-zero. Yet if E is all we have on offer (completed by whatever tacit premises we accept), then E must be conditionally true, thus it is irrelevant how likely or surprising D is. If $Pr(D|E) = 0$ (precisely 0) then, given D, E must be false even if E is the only possible explanation. In cases like that, we are left in ignorance, which is no bad thing, as long as we admit it.

How did we arrive at E anyway? Well, we searched our past experiences and hit upon the explanation which would make the probability of D, given that explanation E, high, if not certain. Since all we could come up with is this E, then, conditionally on our experiences, this E is the best—the only—explanation.

This is one of those areas where notation is both responsible and is the fix for our troubles. There isn't anything wrong with Bayes's theorem (how could there be?), but only with improper uses of it like in (7.1). Pr(E) isn't the "prior" probability of the explanation "before" we see D; it isn't anything; it is a set of meaningless symbols. To become meaningful, some premises are needed. Perhaps "Experience":

$$Pr(E|D, \text{Experience}) = \frac{Pr(D|E, \text{Experience})\,Pr(E|\text{Experience})}{Pr(D|\text{Experience})}. \tag{7.2}$$

And this works if our experience suggests only this explanation because

$$Pr(D|E, \text{Experience}) = Pr(D|\text{Experience}),$$

neither of which are 0, and because $Pr(E|\text{Experience}) = 1$. That makes the left hand side 1, too. No magic has been done here, and it is as simple as said above. Because this was the only E we could think of, this is the only possible explanation; a tautology. Note carefully that probability *qua* probability, nor Bayes's theorem, tells us *how* we came to E, what *caused* us to think of this explanation. That is done by induction, as was shown earlier. Not the for the last time I emphasize probability does not solve all problems of uncertainty and evidential reasoning.

Instead of only one possibility, suppose our intuition or experience suggested two possibilities—or we are told to consider only two possibilities, which amounts to the same thing—E_1= "Garage painting with splatter," E_2= "Scamps drove by and paint-bombed dad." Experience, or outside directive or whatever, is the proposition "Just one of E_1 or E_2 is the correct explanation." (This is similar to our familiar two-sided coin proposition and is not a tautology.) We then have

$$Pr(E_1|D, \text{Experience}) = \frac{Pr(D|E_1, \text{Experience})\, Pr(E_1|\text{Experience})}{Pr(D|\text{Experience})}. \qquad (7.3)$$

Now

$$Pr(E_1|\text{Experience}) = Pr(E_1|E_1 \text{ or } E_2) = 1/2,$$

and

$$Pr(D|E_1, E_1 \text{ or } E_2) = Pr(D|E_1).$$

And it must be that $Pr(D|\text{Experience}) = Pr(D|E_1 \text{ or } E_2) = 1$ since at least one of these must explain the data we have, and *to explain* means to describe the cause. That means $Pr(E_1|D, \text{Experience}) = 0.5 \times Pr(D|E_1)$ and thus $Pr(E_2|D, \text{Experience}) = 0.5 \times Pr(D|E_2)$.

But wait: if E_1 is an *explanation*, then $Pr(D|E_1) = 1$, which E_1 says why D came out; and the same is true for E_2! That makes $Pr(E_1|D, \text{Experience}) = Pr(E_2|D, \text{Experience}) = 1/2$.

And this should make eminent sense: we start with the premise that one of E_1 or E_2 *must* be the cause or explanation of D, which is a trivial premise. Since either can explain D, either has equi-probability of being the cause. There is *nothing* in D to help us pick: D is a consequence of either explanation. It is now clear that inference to the best explanation is a misnomer in any sense which makes use of an observation. There was never a problem with the non-illuminating mathematics. It was our misapplication, writing things like $Pr(D)$, that caused confusion.

Yet it seems in the context of paint-splattered father that his painting the garage is more likely than his being paint-bombed. Why? Both activities would cause the same paint; and our intuitions or some outside directives told us only to consider these and no other explanations. Well, at least my experience—my premises, which are rich and not all articulable—suggests that paint-bombing is pretty rare and garage painting isn't. If I had to pick either E_1 or E_2 given these premises and the deduced non-numerical probability, that is, of course, to make a decision. And that means accounting for non-probability matters like loses and gains: probability isn't decision. Supposing none of those are important, and using the (not here justified rule) of picking the most likely cause, I'd pick garage painting. Simple as that. Once again, to understand cause and explanation, is to seek after knowledge of nature and essence.

A word about the multiplicity of cause. Olympic runners competing for the 100 m run. The gold medalist comes in at 9.69 s, while the silver medalist is right behind at 9.67 s. The winner has won by being faster. We do not need a statistical test to verify this, assuming (conditional on) the measurement equipment is working properly— but then we *never* need a statistical test or model to tell us what happened in the absence of measurement error. Something caused the winner to win. The cause was not one thing, but many, and this is because the movement of the human body over that distance comes about not by one cause but by many. Some of these causes will be more important than others, and these can (possibly) be known by controlling and measuring whatever these conditions are. The exact number of causes is enormous and scarcely countable: among these causes are everything the athlete has eaten (over such-and-such a time), all the elements that enter into his training regime, and on and on.

Next consider winning times over a number of races. Men gold medalists are always faster than the fastest woman. Does male sex cause the men to out-race their feminine competitors? As everybody knows, it is not sex *per se* that causes men to be faster; instead, sex causes differences in anatomy and physiology that are tied to athletic performance. Men have more muscle, and muscle causes fleetness, and so on. This is why the myriad of models that "control" for sex and which imply sex is a cause are always wrong—unless they are modeling direct effects of sex, such as pregnancy, and in which case, no model is used because we understand the essence. Sex is a proxy for (usually) multiple other causes and is itself not a cause. And this kind of reasoning also applies for such things as race, income, and so on. Statistical models aren't capable of discerning cause.

7.7 Under-Determination

Per the Duhem-Quine thesis, scientific theories are under-determined and through any finite collection of facts you can always draw multiple theories, e.g. [135, 150, 153, 179, 180]. And explanatory power is not a guarantee a true cause has been discovered. A contemporary political example might be best in displaying the difficulty under-determination poses for discovering causes because of the importance people place on social questions. Chess Grandmaster Nigel Short said publicly that men and women are different and that men are better at chess than women. He said this was be*cause* the two sexes are "hard-wired very differently". According to a newspaper account [218], Short said:

Why should they [men and women] function in the same way? I don't have the slightest problem in acknowledging that my wife possesses a much higher degree of emotional intelligence than I do...we just have different skills. It would be wonderful to see more girls playing chess, and at a higher level, but rather than fretting about inequality, perhaps we should just gracefully accept it as a fact.

A chess player and writer named Amanda Ross said in response to Short that it is "incredibly damaging when someone so respected basically endorses sexism". Without being especially careful, we can define *sexism* as when a disparity of some kind exists between males and females and which is caused by men or society itself "pushing down" or otherwise limiting females. Short argues for biological differences; Ross claims egalitarianism is true. Now Ross begins her argument with the observation that a female once beat Short in a game, which to Ross proves males and females are equal. And she would be right if "equal at chess" meant "some woman somewhere can beat some man at chess." Evidently, this is not what Short meant when he said men and women were "unequal at chess." He meant something like, "In any list of top players, the majority will be men."

There is a list of Grandmasters, the highest title chess players can earn (awarded by the World Chess Federation, [221]). As of this writing, there are 1413 men and 33 females on the list. This evidence bolsters Short's claim that men are better in the sense he meant. But it *also* boosts Ross's theory that sexism is rampant. The same data supports *both* theories. Short says men are superior chess players and here is a list showing they are. But Ross says mankind (and presumably culture) is sexist and keeps women from reaching top levels, and here is a list showing her prediction is right. The data cannot decide which theory is true. The theories are under-determined. And, of course, other theories might also explain the data. Men and women might be in essence equally skilled, but men like playing more. Or are allowed to. And so on.

There is no use bringing in Bayes's Theorem and asking about "prior" probabilities on the truth of each theory, because the holders of both theories start by believing they are true. The data we have can't shake either Short or Ross free from the conviction he or she is right. The data wouldn't help us (from our perspective) either if we are indifferent between the theories. It *is* true that different data might support Short and Ross differently. Suppose on the list were 722 men and 722 non-men. Ross, claiming the triumph of equality over sexism, is upheld. But then it would be Short's turn to claim that men really are superior, but the culture is pushing *them* down.

There is no general solution. The under-determination of the contingent is a fact. Collecting more data wouldn't work, either. What we have to do, and even this is not a complete solution as is by now clear, is to look *outside* the data. For instance, we might argue that chess is an abstract analytical activity. If Short is right, men should be better than women at other abstract analytical activities. What's more analytical than mathematics? The list of Fields medalists contains only one woman.

For Short to be right, only one thing must be true: men must have different brains, i.e. the essence of men must be different than the essence of women. For Ross to be right, many more things must be happening in more places and at more times. Sexism under Ross's scheme must operate like the nervous say the Trilateral Commission does: a worldwide occult top secret network with strange unstoppable powers. If we accept the premise that fewer premises are more often associated with true theories, then there is good evidence Short is right. But it's doubtful we'd get Ross to agree. Or perhaps some of you have thought of other objections to Ross; just as some of you might have thought of criticisms of Short. Jaynes's [122, Chapter 5] on the Queer Uses of Probability is not to be missed on this subject; Jaynes has several numerical examples that will be of interest.

Chapter 8
Probability Models

"But what a weak barrier is truth when it stands in the way of an hypothesis."—Mary Wollstonecraft Shelley.

A model is an argument. Models are collections of various premises which we assign to an observable proposition (or just "observable"). That is, modelling reverses the probability equation: the proposition of interest or conclusion, i.e. the observable Y, is specified first after which premises X thought probative of the observable are sought or discovered. The ultimate goal is to discover just those premises X which cause or which determine Y. Absent these—and there may be many causes of Y—it is hoped to find X which give Y probabilities close to 0 or 1, given X in its various states.

Not all probability models apply to observable Y. But an implicit premise to all observables is that Y is contingent. Given just that premise, (as we have seen) the probability of Y is the unit interval *sans* endpoints. Models which supply X which sharpen this probability are of potential interest. The more the probabilities are sharpened the more interesting the model. Interestingness and usefulness are not identical. A model's usefulness is described by what decisions are made with it, and how costly and how rewarding those decisions are. When calculations of usefulness are possible, which is rare for "public models" (such as those which appear in academic journals) except by gross approximation, usefulness is reckoned by a *proper score*. The usefulness of models are easily compared by scores; when one model has a better score than another, the superior model is said to have skill with respect to the lesser (conditional on the score type). Only proper scores should be used.

A probability model is the same as a causal or deterministic model except that the propositions of interest Y are not all certainly true or false given the collection of premises X. Our old friend, given X = "This is a two-sided object which when flipped must show one of H or T" the proposition Y = "An H shows" is neither true nor false, but in-between. There needn't be a real object which conforms to these premises, although many can. Do not forget the empirical bias in discussions

© Springer International Publishing Switzerland 2016
W. Briggs, *Uncertainty*, DOI 10.1007/978-3-319-39756-6_8

of probability. To implement this "coin" model in real life, simply find any two-sided or two-state object which conforms to the premises or something like them. This becomes a model. Whether this model is useful for this object is a different question.

Probability models can and do have causative elements. But these are generally found in very large, integrated physical models, such as weather and climate models, engineering models which investigate air or water flow over vessels, and that sort of thing. These are models which might even be fully causal or deterministic in the sense given last chapter, but which are treated as probabilistic in practice. Tacit premises are added to the predictions from these models which adds uncertainty so that falsification is avoided.

8.1 Model Form

Recalling our goal is not the particulars of a mathematical theory but of understanding, a probability model takes the form:

$$\Pr(Y|X), \tag{8.1}$$

where, as is usual by now, Y is the proposition of interest and X the compound list of premises thought probative of Y. A probability model produces non-extreme probabilities for Y for the Xs considered, whereas a causal or deterministic model produces only extreme, i.e. 0–1, probabilities for all X considered. Example: Given X = "A Metalunan interocitor must be in one of n states, $s_1, s_2, \ldots, s_n$ and S is a interocitor" to discover the probability of Y = "S is in state s_j". Statistical and physical models are, as we learn next chapter, probability models applied to observable Y. Probability models are thus more general than statistical models.

Models are often written with parameterized Xs in forms like this:

$$Y \sim \mathrm{N}(m, s^2), \tag{8.2}$$

where Y becomes variable or changeable. A loose example: Y is Y = "The value is y_1", "The value is y_2", etc. This is a sloppy way of writing the model because the X (the evidence or premise) which gave rise to the normal is missing; only the parameters and "distribution" remains, as if falling from above. This form also tends to reify the probability, making it appear that Y is caused, made *to be*, by the normal distribution, which is impossible. This form is why things like this are often (always?) heard: "Temperature is normally distributed". No it is *not*. Temperature is caused. Our uncertainty in the value temperature might take is quantified (but only ever approximately) by a normal distribution with central parameter m and spread s; in other words, given some X which specifies these values. Incidentally, for unfortunate historical reasons, these parameters often take for names their classical estimates, "mean" and "variance". That seemingly small error leads to pervasive over-certainty when m and s are unknown; and they are usually are not known.

If it's not already obvious, (8.1) is written in a predictive form, saying something about Y which we do not already know, or do not choose to know; anyway, Y itself is not part of X. If it were, the model would be circular and produce extreme probabilities, which would make it causal or deterministic. Equation (8.2) is also written predictively, though it might not seem like it is. But that is because attention in classical statistics takes inordinate interest in the parameters and not on the model *per se*. The limitations of that approach are discussed below and in the next chapter.

The reader should recall the discussion of the relation between probability models and causality. If we knew (somehow) X was *not* causally or deterministically related to Y, then X should *never* be used in a model of Y. Any association of X with Y in these cases is coincidental, spurious. It is only if we *accept* that X is causally or deterministically related to Y that it should appear in a model of the uncertainty of Y. Incidentally, it is a model of the *uncertainty* of Y, and *not* a model of Y; the latter language is casual or deterministic. That should be repeated, and even shouted: *these are not models* of *Y but of the uncertainty* in *Y!* Only causal and deterministic models are models *of* Y. Probability models, don't forget, rest on the assumption of causal or deterministic relations somewhere in the universe, and in uncertainty remains, it must be that *blocking* of these causes must be occurring at least sometimes.

8.2 Relevance and Importance

As covered in Chap. 4, an X that is added to a model which, in the presence of the other premises (other "Xs"), does not change the probability of Y is irrelevant or not probative. The data X which are used to "fit" a model are of course themselves premises, i.e. $X_{1,1} =$ "Observed the value 112.2 mm/Hg" (for the first premise, say of systolic blood pressure, first observation from some collection). The importance of each premise, given the presence of the other premises, is judged by how much it changes the probability of Y. If an X does not result in extreme probabilities of Y, this X is not necessarily causal, though an injurious, flabbergasting tradition has developed (predominately in the "soft" sciences) which says or assumes it is.

For example, if $Pr(Y|X_1) = Pr(Y|X_1X_2)$ then X_2 is irrelevant in the presence of X_1, even if $Pr(Y|X_2)$ is something other than the unit interval. That is, X_2 may be separately probative of Y but it adds no information about Y that is not already in X_1. There are thus two kinds of relevance, *in-model*, which is rather a measure of importance, how much a premise changes our understanding of Y, and *out-model*, whether the premises is even needed. A third is a variant of the first: *sample* relevance.

Suppose Y itself takes different states (like temperature) and that $Pr(Y_a|X_1) = Pr(Y_a|X_1X_2)$ but $Pr(Y_b|X_1) \neq Pr(Y_b|X_1X_2)$. X_2 in the presence of X_1—the condition which must *always* be stated—is then relevant to Y; or, better, relevant only when Y is Y_b.

Suppose $\Pr(Y|X_1) = \Pr(Y|X_1X_2) + \epsilon$, with the obvious constraints on ϵ. Then X_2 in the presence of X_1 is relevant. Whether the difference ϵ makes any difference to any decision is not a question probability can answer. "Practically irrelevant" is *not* irrelevant. Irrelevance is a logical condition. The practice of modeling is the art of selecting those X (premises) which are relevant, in the presence of other premises, to Y. Invariably, some new premise will add "only" ϵ to the understanding of Y. Whether this is enough to "make a difference" is a question only the modeler and whomever is going to use the model can answer. The *only* "test" for relevance is thus *any* change in the conditional probability of Y.

Relevance, as we see next chapter, is how models should be judged before verification of their predictions arrive. Assessing relevance is hard work—but who said modeling had to be easy? That modeling is now far too easy is a major problem; because anybody can do it, everybody thinks they're good at it. Supposing Y is simple (yes or no, true or false), and a list of premises, the relevance of each X_i— its subscript indicates it is variable—is assessed by holding those other X_j which are variable at some fixed level and then varying the X_i. For example, to assess the relevance of X_1, which can take the values a_1 and a_2, compute

$$\Pr(Y|X_1 = a_1, X_2 = b, \ldots, X_p = p, W),$$

where W are those premises which are fixed (deductions, assumptions, etc.), and

$$\Pr(Y|X_1 = a_2, X_2 = b, \ldots, X_p = p, W).$$

The difference between these two probabilities is the *in-model* relevance of X_1 given the values the other X take. The *out-model* relevance is assessed by next computing

$$\Pr(Y|X_2 = b, \ldots, X_p = p, W),$$

and comparing that to the model which retains X_1. Note that all the other X have kept their values. *Sample* relevance is computed by calculating the same probability but or the addition (or subtraction) of a new "data point." Irrelevance is:

$$\Pr(Y|X_{n+1}) = \Pr(Y|X_n).$$

For instance, suppose we have n observations and on the $n + 1$ the probability Y is true remains unchanged. Then this new data point has added no new information, *in the presence of* the first n. Of course, these may be used to hunt for those data points which are most relevant, or rather, most important, and which are irrelevant (given the others). Those familiar with classical parametric methods will see the similarities; this approach is superior because all measures are stated directly and with respect to the proposition of interest Y.

I should highlight I am not here trying to develop a set of procedures *per se*, only defining the philosophically relevant constituents of probability models. We want to know what it means to be a probability model—*any* probability model, and not just one for some stated purpose. Readers interested in working on new problems will discover lots of fertile ground here, though.

It should be by now obvious that each of these probabilities is also a *prediction*. They say "Here is the probability of Y should all these other things hold." So not only probabilities, but all predictions are conditional, too. This form of model also forms the basis of how statistical methods should work. All concentration is centered on asking how X influences our knowledge of Y—and even in rare cases, how X causes or determines Y.

Relevance is made more difficult when Y is allowed to vary, but the underlying idea is the same. Except for the X of interest which is varied or removed from the model, fix the others Xs, and compute the probability of the Ys and see what changes to this probability happen. Relevance is when there are changes, and irrelevance when not. This is obviously going to be a lot of work for complex Ys and Xs, but nothing else gives a fairer and more complete picture of the uncertainty inherent in the problem. And, again, who said it had to be easy?

8.3 Independence Versus Irrelevance

What's the difference between *independence* and *irrelevance* and why does that difference matter? This typical passage from *The First Course in Probability* by Sheldon Ross [185, p. 87] is lovely because many major misunderstandings are committed, all of which prove "independence" a poor term. And this is a book I highly recommend as an introduction to probability calculations; for readability, I changed Ross's notation slightly, from e.g. "$P(E)$" to "$\Pr(E)$" to keep in the style of this book.

> The previous examples of this chapter show that $\Pr(E|F)$, the conditional probability of E given F, is not generally equal to $\Pr(E)$, the unconditional probability of E. In other words, knowing that F has occurred generally changes the chances of E's occurrence. In the special cases where $\Pr(E|F)$ does in fact equal $\Pr(E)$, we say that E is independent of F. That is, E is independent of F if knowledge that F occurred does not change the probability that E occurs.
> Since $\Pr(E|F) = \Pr(EF)/\Pr(F)$, we see that E is independent of F if $\Pr(EF) = \Pr(E)\Pr(F)$.

The first misunderstanding is "$\Pr(E)$, the unconditional probability of E". There is no such thing. No unconditional probability exists, as shown earlier. All, each, every probability must be conditioned on something, some premise, some evidence, some belief. Writing probabilities like "$\Pr(E)$" is always, every time, an error, not only of notation but of thinking. It encourages and amplifies the false belief that probability is a physical, tangible, implicit, measurable thing. It also heightens the second misunderstanding. We must always write (say) $\Pr(E|X)$, where X is whatever evidence one has in mind.

The second misunderstanding, albeit minor, is this: "knowing that F has occurred generally changes the chances of E's occurrence." Note the bias towards empiricism. We do not have to deal with observables in probability models, though we do in statistical and physical. In other places Ross writes, in order to judge these probabilities, we must imagine "An infinite sequence of independent trials is to be performed" (p. 90), which is an impossibility. Another misconception: "Independent trials, consisting of rolling a pair of fair dice, are performed (p. 92). We already learned "fair" dice are impossible in practice. "Events" or "trials" "occur", says Ross, echoing many other authors, which are propositions that can be measured in reality, or are mistakenly thought to be measurable. Probability is much richer than that and applies to propositions that are not observable.

Non-empirical propositions, as in logic, easily have probabilities, as we recall. Example: the probability of E = "A winged horse is picked" given X = "One of a winged horse or a one-eyed one-horned flying purple people eater must be picked" is 1/2, despite that "events" E and X will never occur. So maybe the misunderstanding, or the empirical bias, isn't so minor at that. The bias towards empiricism is what partly accounts for the frequentist fallacy, about which we already know something; but there is more to say below. Notice that our example E and X have *no* limiting relative frequency. Instead, we should say of any Pr(E|F), "The probability of E (being true) accepting F (is true)."

Those missteps are common and not the main difficulty. The third and grand-daddy of all misunderstandings is this: "E is independent of F if knowledge that F occurred does not change the probability that E occurs." The misunderstanding comes in two parts: (1) use of "independence", and (2) a mistaken calculation.

Number (2) first. It is a mistake to write "Pr(EF) = Pr(E) Pr(F)" because given the same E and F, there are times when this equation holds and times when it doesn't. A simple example. Let E = "The score of the game is greater than or equal to 4" and F = "Device one shows 2". What is Pr(E|F)? Impossible to say: we have no evidence tying the device to the game. Similarly, Pr(E) does not exist, nor does Pr(F).

Let X = "The game is scored by adding the total on devices one and two, where each device can show the numbers 1 through 6." Then $\Pr(E|X) = 33/36, \Pr(F|X) = 1/6$, and $\Pr(E|FX) = 5/6$; thus $\Pr(E|X)\Pr(F|X)$ (~ 0.153) which does not equal $\Pr(E|FX)\Pr(F|X)$ (~ 0.139). Knowledge of F *in the face of X* is *relevant* to the probability E is true. Recall these do not have to be real devices; *they can be entirely imaginary.*

Now let W = "The game is scored by the number shown on device two, where device one and two can show the numbers 1 through 6." Then $\Pr(E|W) = 1/2, \Pr(F|W) = 1/6$, and $\Pr(E|FW) = 1/2$ because knowledge of F *in the face of W* is *irrelevant* to knowledge of E. In this case $\Pr(EF|W) = \Pr(E|W)\Pr(F|W)$.

The key, as might have always been obvious, is that relevance depends on the specific information one supposes.

Number (1). Use of "independent" conjures up images of causation, as if through dependence, somehow, F is causing, or causing something which is causing, E. This error often happens in discussions of time series, as if previous time points caused current ones. We have all heard times without number people say things like, "You can't use that model because the events aren't independent." But you can use any model you like, it's only that some models make better use of information because, usually, knowing what came before is *relevant* to predictions of what will come. Probability is a measure of information, not a quantification of cause.

Here is another example from Ross showing this misunderstanding (p. 88, where the author manages two digs at his political enemies):

> If we let E denote the event that the next president is a Republican and F the event that there will be a major earthquake within the next year, then most people would probably be willing to assume E an F are independent. However, there would probably be some controversy over whether it is reasonable to assume that E is independent of G, where G is the event that there will be a recession within two years after the election.

To understand the second example, recall that Ross was writing at a time when it was still possible to distinguish between Republicans and Democrats. The idea that F or G are the full or partial efficient cause of E suffuses this example, a mistake reinforced by using the word "independence". If instead we say that knowledge of the president's party is *irrelevant* to predicting whether an earthquake will soon occur we make more sense. The same is true if we say knowledge that this president's policies are relevant for guessing whether a recession will occur.

This classic example is a cliché, but is apt. Ice cream sales, we hear, are positively correlated with drownings. The two events, a statistician might say, are not "independent". Yet it's not the ice cream that is *causing* the drownings. Still, *knowledge* that more ice cream being sold is *relevant* to fixing a probability more drownings will be seen! The *model* is still good even though it is silent on cause. This point cannot be stressed too highly. Good and useful models can badly screw up causes but they can still make useful predictions. A woman can insist gremlins power her automobile and still get where she's going.

The distinction between "independence" and "irrelevance" was first made by Keynes in his unjustly neglected *A Treatise on Probability* [132, pp. 59–61]. Keynes argued for the latter term, correctly asserting, first, that no probabilities are unconditional. Keynes gives two definitions of irrelevance, which amplify the previous section. In my notation but his words, "F is irrelevant to E on evidence X, if the probability of E on evidence FX is the same as its probability on evidence X; i.e. F is irrelevant to E|X if $Pr(E|FX) = Pr(E|X)$". This is as above.

Keynes tightens this to a second definition. "F is irrelevant to E on evidence X, if there is no proposition, inferrible from FX but not from X, such that its addition to evidence X affects the probability of E." In our notation, "F is irrelevant to E|X, if there is no proposition F' such that $Pr(F'|FX) = 1$, $Pr(F'|X) \neq 1$, and $Pr(E|F'X) \neq Pr(E|X)$." Note that Keynes has kept the logical distinction throughout ("inferrible from"). Lastly, Keynes introduces another distinction (p. 60):

> h_1 and h_2 are independent and complementary parts of the evidence, if between them they make up h and neither can be inferred from the other. If x is the conclusion, and h_1 and h_2 are independent and complementary parts of the evidence, then h_1 is relevant if the addition of it to h_2 affects the probability of x.

This passage has the pertinent footnote (in my modified notation): "I.e. (in symbolism) h_1 and h_2 are independent and complementary parts of h if $h_1 h_2 = h$, $Pr(h_1|h_2) \neq 1$, and $\Pr(h_2|h_1) \neq 1$. Also h_1 is relevant if $Pr(x|h) \neq Pr(x|h_2)$."

Keynes's formulation emphasizes that it is not only the "raw" X which are premises, but those propositions which can be deduced from them, which was mentioned above but not emphasized. Note: two (or however many) observed data points, say, x_1 and x_2 are independent and complementary parts of the evidence because neither can be deduced—not mathematically or logically derived *per se*—from each other. Observations are thus no different than any other proposition. In other words, every observation if of the schema, X = "The value x was seen".

8.4 Bayes

Bayesian theory isn't what most think. Most believe that it's about "prior beliefs" and "updating" probabilities, or perhaps a way of encapsulating "feelings" quantitatively. The real innovation is something much more profound. And really, when it comes down to it, Bayes's theorem isn't even necessary for Bayesian theory. Here's why.

Again, any probability is denoted by the schematic equation $\Pr(Y|X)$, which is the probability the proposition Y is true given the premise X. As always, X may be compound, complex or simple. Bayes's theorem looks like this:

$$\Pr(Y|WX) = \frac{\Pr(W|YX)\,\Pr(Y|X)}{\Pr(W|X)}. \tag{8.3}$$

We start knowing or accepting the premise X, then later assume or learn W, and are able to calculate, or "update", the probability of Y given this new information. Bayes's theorem is a way to compute $\Pr(Y|WX)$. But it isn't strictly needed. We could compute $\Pr(Y|WX)$ directly from knowledge of W and X themselves. Sometimes the use of Bayes's theorem can hinder.

An example. Given X = "This machine must take one of states S_1, S_2, or S_3" we want the probability Y = "The machine is in state S_1." The answer is 1/3. We then learn W = "The machine is malfunctioning and cannot take state S_3". The probability of Y given W and X is 1/2, as is trivial to see. Now find the result by applying Bayes's theorem, the results of which must match. We know that $\Pr(W|YX)/\Pr(W|X) = 3/2$ (because $\Pr(Y|X) = 1/3$). But it's difficult at first to tell how this comes about. What exactly is $\Pr(W|X)$, the probability the machine malfunctions such that it cannot take state S_3 given only the knowledge that it must take one of S_1, S_2, or S_3? If we argue that if the machine is going to malfunction, given the premises we have (X), it is equally likely to be any of the three states, thus the probability is 1/3. Then $\Pr(W|YX)$ must equal 1/2, but why? Given we know the machine is in state S_1, and that it can take any of the three, the probability state S_3 is the malfunction *is* 1/2, because we know the malfunctioning state cannot be S_1, but can be S_2 or S_3. Using Bayes works, as it must, but in this case it added considerably to the burden of the calculation.

Most scientific, which is to say empirical, propositions start with the premise that they are contingent. This knowledge is usually left tacit; it rarely (or never) appears in equations. But it could: we could compute $\Pr(Y|Y$ is contingent), which even is quantifiable (the open interval $(0, 1)$). We then "update" this to $\Pr(Y|X \& Y$ is contingent), which is 1/3 as above (students should derive this). Bayes's theorem is again not needed.

Of course, there are many instances in which Bayes facilitates. Without this tool we would be more than hard pressed to calculate many probabilities. But the point is the theorem can but doesn't have to be invoked as a computational aide. The theorem is not the philosophy.

The real innovation in Bayesian philosophy, whether it is recognized or not, is the idea that any uncertain proposition can and must be assigned a probability. This dictum is not always assiduously followed; and the assignment need not be numerical. This is contrasted with frequentist theory which assigns probabilities to some unknown propositions while forbidding this assignment in others, and where the choice is *ad hoc*. Bayesian theory has two main flavors, subjective and objective. The subjective branch assigns probabilities based on emotions or "feelings", a practice we earlier saw leads to absurdities. Objective theory tends to insist every probability can and should be quantified, which also leads to mistakes. We'll find a third path by quantifying only that which can be quantified and by not making numbers up to satisfy our mathematical urges.

8.5 The Problem and Origin of Parameters

There has been an inordinate and unfortunate fascination with unobservable parameters which are found inside most probability models. Parameters relate the X to Y, but are understood in an *ad hoc* fashion; see [210, 212, 213]. Since models are often selected through custom or ignorance of alternatives (and recall we're talking about actual and not ideal practice), the purposes of parameters are not well considered, to say the least. Most statistical practice, frequentist or Bayesian, revolves solely around parameters, which has led to the harmful misconception that the parameters are themselves the X, and the X causal. *P*-values, confidence intervals, and posterior distributions, hypothesis tests and other classic measures of model "fit" are abused with shocking frequency and with destructive force. Probability leakage is the least of these problems; mis-ascribed causality the worst. It's time for it to stop. People want to know $\Pr(Y|X)$: tell them *that* and not about some mysterious parameters.

Parameters arise from considering measurement. All measurement is finite and discrete, regardless of the way the universe might happen to be (I use *universe* in the philosophical sense of *all that exists*). Measurement drives X, which in turn are probative of Y. Parameters are not necessary when all is finite and discrete, but they may be used for mathematical convenience. But their origin must first be understood. Parameters-as-approximations arise from taking a finite discrete measurement process (which all are processes) to the limit. The interpretation of parameters in this context then becomes natural. This area, as will soon be clear, is wide open for research. Below, I'll show how parameters arise in a familiar set up; but how they come about in others is mostly an open question.

Where do parameters come from? Here is one example, which originates with Laplace, and which necessitates some mathematics, which, I remind us, are not our main purpose here. The parallels to Solomonoff's approach (cited in Chap. 6) will be obvious to those familiar with algorithmic information theory. Begin with the premise E that before us in an urn which contains N objects, objects which can take one of two states. From this language we infer N is finite, which is absolutely no restriction, because N can be very large indeed. Call them "success" and "failure", or "1" and "0", if you like. From this we deduce there can be anywhere from 0 to N successes. Given these premises—and *no* others—or rather this evidence E, we deduce the probability that there are $M = i, i = 0, \ldots, N$ successes is $1/(N+1)$. No number of success (or failures) is more likely than any other.

Now suppose we reach in and grab a sample of size n. In this sample there will be n_1 success and n_0 failures, so that $n_1 + n_0 = n$. To say something about these observations, we want the probability of j successes in n draws, without replacement where the urn has N total successes. It will also be helpful to rewrite, or rather parameterize this by considering $N\theta$, where $\theta = i/N$, which is the fraction of successes. Note that θ is observable. The probability is (with the obvious restrictions on j):

$$\Pr(n_1 = j | n, \theta, N, E) = \frac{\binom{N\theta}{j}\binom{N-N\theta}{n-j}}{\binom{N}{n}}, \qquad (8.4)$$

which is a hypergeometric. We are still interested in the fraction θ (out of all N) of successes. Since we saw n_1 successes so far, θ must be at least as large as n_1/N, but it might be larger. We can use Bayes's theorem to write (again, with the obvious restrictions on j)

$$\Pr(\theta = j/N | n, n_1, N, E) \propto \Pr(n_1 = j | n, \theta = j/N, N, E) \Pr(\theta = j | n, N, E). \qquad (8.5)$$

This is the posterior "parameter" distribution on θ, which turns out to be

$$\Pr(\theta = j/N | n, n_1, N, E) = \binom{N-n}{j-n_1} \frac{\beta(j+1, N-j+1)}{(n+1)\beta(n_1+1, n_0+1)}, \qquad (8.6)$$

where $\beta()$ denotes the beta function.

Here is where parameters arise. Following Ross (*op cit.*, p. 180) in showing how the hypergeometric is related to the binomial for large samples, let $N \to \infty$ in (8.6). The result is

$$\lim_{N \to \infty} \frac{1}{N+1} \frac{\binom{N\theta}{j}\binom{N-N\theta}{n-j}}{\binom{N}{n}} = \binom{n}{n_1} \theta^{(n_1+1)-1}(1-\theta)^{(n_0+1)-1}, \tag{8.7}$$

which is the standard beta distribution posterior on θ when the prior on θ is "flat", i.e. equal to a beta distribution with parameters $\alpha = \beta = 1$.

We started with hard-and-fast observable propositions, and a finite number of successes and failures, and considered how expanding their number *in a specific way* towards infinity, and we end up with unobservable parameters. As Jack Aubrey would say, Ain't you amazed? The key is that we don't really need the infinite version of the model; the finite one worked just fine, albeit that it is harder to calculate for large N. But then there is no arguments over where the prior for θ came from. It arises naturally. This small demonstration is like de Finetti's representation theorem (see below), only it also gives the prior instead of saying only that it exists.

What does the parameter θ mean? With a finite N—which will always be true of all real-world situations—it was the total fraction of successes (given the premises). This is sensible and measurable, at least in theory. Whether anybody ever measures all N mentioned in the premises is another matter. θ is discrete: it can take *only* the values $0/N, 1/N, \ldots, N/N$, and no value inside this set is impossible; at least, not on the evidence we have assumed. At the limit, θ is continuous and can take any value in the unit interval. Which is to say, it can take none of them, not empirically, because as Keynes said, in the long run we shall all be dead: infinity can never be reached. The parameter is no longer the fraction of successes, only something like it. But what? The mind should boggle at imagining the ratio infinite successes in infinite "chances"; indeed, I cannot imagine it. I can only picture large finite approximations to it. This θ is *not*, as it is often called, "the probability of success." We already deduced the probability of success given the premises. So what is it? An index with a complex definition involving limits, a definition so complex that its niceties are forgotten and people speak of it as if it were its finite cousin, that is, as it if were a probability.

Notice very carefully that the parameter-solution is an *approximation*. We don't need it. Though calculating (8.6) may be cumbersome, we have the exact result. We don't need to quarrel about the prior, impropriety, ignorance, non-informativity or anything else because everything has been *deduced* from the premises. This situation is also well behaved. Approaching the limit (in a certain specified way) produced a result which is familiar. The continuous-valued parameter ties nicely to a finite-sample result: it keeps the roughly same meaning. I have no idea whether this will be true for all stock distributions in our cookbook, but we have great reason to doubt it. In his book, Jaynes (*op cit.*, Chapter 15) shows how the so-called marginalization paradox disappears when one very carefully tracks how one heads off to infinity. Buffon's needle paradox is another well known example where the path matters.

There's more to this example. Suppose we've taken our sample, n_0 and n_1, and want to know the possibilities for the remaining number of successes d. This is $d = N\theta - n_1$, so $N\theta = d + n_1$ (N is finite here). Then

$$\Pr(D = d | n, n_1, N, \mathrm{E}) \propto \frac{1}{n+1} \frac{\binom{n_1+d}{d}\binom{N-n_1-d}{N-n-d}}{\binom{N+1}{n+1}}. \tag{8.8}$$

This is an unnormalized form of the negative hypergeometric, which is an alternate name for the beta-binomial distribution. Before we have removed any of the balls from the urn, we can compute the prior predictive distribution of n_1 in the initial sample. This quantity is critical since it forms the normalization constant of the posterior distribution in Eq. (8.6). To derive this we simply sum over the possible values of θ in expression (8.8) which is equivalent to summing a beta-binomial mass function over its range. The result is

$$\Pr(n_1 = j | n, N) = 1/(n+1), \qquad j \in \{0, \ldots n\}. \tag{8.9}$$

If we reach in and grab out just one ball, the chance that it is a 1 or 0 is 1/2, no matter what N is. This result is well known and forms the basis of Laplace's rule of succession. Furthermore, if we grab $n > 1$ balls, the result says that n_1 is equally likely to be any result in $\{0, 1, \ldots, n\}$, and where knowledge of N is also irrelevant. This is intuitive, since we began with all proportions θ being equally likely *a priori* and have not yet collected data that suggests otherwise.

Given the original sample A whose size is now denoted by n_a, with n_{1a} the number of successes and n_{0a} the number of failures, let a new sample of size $n_b \leq N - n_a$ be collected. We want to know the distribution of successes, n_{1b}, in this new sample. It is known that $0 \leq n_{1b} < N\theta - n_a$. Given d, the distribution of n_{1b} is clearly hypergeometric.

This allows us to compute the posterior predictive distribution on n_{1b}. This turns out to be:

$$\Pr(n_{1b} | n_b, n_{1a}, n_a, N, \mathrm{E}) = \binom{n_b}{n_{1b}} \frac{\beta(n_{1a} + n_{1b} + 1, n_{0a} + n_{0b} + 1)}{\beta(n_{1a} + 1, n_{0a} + 1)} \tag{8.10}$$

which is a beta-binomial distribution with parameters $(n_b, n_{a1} + 1, n_{a0} + 1)$. Knowledge of N is irrelevant here, too, except in the weak sense the total sample $n_a + n_b \leq N$. Also, θ does not appear. This quantity is far more interesting that anything we have to say about θ. It conforms to the true goal of statistical modeling, which is to say sensible things about that which can be measured. Interestingly, this is the same answer one gets starting from a "flat" prior on a continuous θ, integrating out the uncertainty in θ and forming the regular posterior predictive distribution.

We did not start with any parameters and we did not end with any. Parameters aren't needed, as promised.

Laplace's rule of succession follows the same course, although he stated it differently; see [228] for another taken on succession. Laplace derived the probability of the next observation in some ill-defined process being a success given we have seen n_{1a} successes in the first n instances. The well known answer is

$$\Pr(x_{n+1} = 1 | n_{1a}, n, \text{fuzzy}) = \frac{n_{1a} + 1}{n + 2} \tag{8.11}$$

where the "fuzzy" indicates Laplace was not quite clear about his premises. It is easy to check that (8.11) is equivalent to (8.10). That means the fuzziness is replaced by our firm premises E. If Laplace had started with E, all would have been well. But he chose an unfortunate example, that of the sun rising. He used the rule of succession to calculate the probability that the sun will rise tomorrow, given that it has risen every day for the past however many years. Call it 6000 years. That makes $n = 2,191,500$ which makes the probability about 0.9999995.

Large, but many objected saying it surely ought to be higher because we know a lot more about sunrises than Laplace admits in his formula. This is true. But it is also beside the point. Laplace's formula is the *right* answer to a different question, that's all. It is a deduced probability given the premises fixed in E. That we apply these premises to a sunrise is our (and Laplace's) mistake. It doesn't make the formula wrong.

Jaynes [122] has an example similar to this, but for the normal, showing the derivation for the normal distribution from premises starting with something like "The error in the measurement can be any value, positive or negative." But a hidden, or rather tacit, premise derivable from that is that the measurement can be continuous, that it has infinite gradations, which is always impossible in practice. Instead, a superior working premise is that that error in measurement can be one of only a set of values, where the values are specified by the apparatus at hand. If this set is allowed to go to the limit, then it is likely (I haven't made the calculations) the normal would result. But if the set is fixed by the apparatus, we don't need to go to the limit. The resulting predictive distribution will be parameter-free, a fully deduced model, just as above (it will resemble, I am guessing, a multinomial in the limit, or something like it in finite form). As I said above, this kind of thing is ripe with research topics.

8.6 Exchangeability and Parameters

Suppose (for simplicity) Y is true-false and is to be assessed more than once. Given E, the usual way to say that Y is *exchangeable* is if

$$\Pr(Y_1 Y_2 \ldots Y_n | E) = \Pr(Y_{\pi_1} Y_{\pi_2} \ldots Y_{\pi_n} | E) \tag{8.12}$$

where π_i is a permutation of the numbers $1, \ldots, n$. The order doesn't matter. Exchangeability is not quite the same as irrelevance (or independence), which might not be obvious given this way of writing. To see why consider a Polya urn model.

We have an urn with n white and m black balls. We grab one out, note its color, and then toss it and another ball of the same color back into the urn. We then grab out a second ball, and repeat. Given this evidence, the probability of grabbing a first white ball is $n/(n + m)$. Suppose the first ball grabbed is white. Another white ball is tossed into the urn, giving now $n + 1$ white balls. Given this updated information, the probability the second ball drawn is white is $(n + 1)/(n + m + 1)$. But then suppose a black ball had been drawn first. Then the probability the second ball is white is $n/(n + m + 1)$. Knowledge of which ball drawn first is relevant to knowing the probability the second ball is white. In other words, the notation above might be incomplete. We really have (with obvious notation)

$$\Pr(\text{first white}|n, m) = \frac{n}{n + m},$$

then

$$\Pr(\text{second white}|n + 1, m) = \frac{n + 1}{n + m + 1},$$

or

$$\Pr(\text{second white}|n, m + 1) = \frac{n}{n + m + 1}.$$

Now, given our evidence, the probability of the first ball being white is, no matter what, $n/(n + m)$. And the probability the second is white is then (simplifying)

$$
\begin{aligned}
\Pr(\text{second white}|E) &= \Pr(w_1w_2 \text{ or } b_1w_2|E) \\
&= \frac{n}{n + m + 1}\frac{n + 1}{n + m + 1} + \frac{m}{n + m + 1}\frac{n}{n + m + 1} \\
&= \frac{n(n + 1) + mn}{n + m + 1} = \frac{n}{n + m}.
\end{aligned}
$$

The probability of white on the first is the same as on the second, as it is on the third or fourth, etcetera. Intuitively this makes sense, because we're augmenting the original urn with more-or-less the same proportion of new white and black balls. But this result is a consequence of the evidence, not the definition of exchangeability. Exchangeability would be if (in this case)

$$\Pr(w_1b_2|E) = \Pr(b_1w_2|E),$$

which is easily seen to be true. Of course, we need not check the cases w_1w_2 and b_1b_2. And it is easily shown that no matter how long the (finite) string is, the sequence is exchangeable, given this evidence. To emphasize, exchangeability is just as conditional as probability and relevance are.

We have earlier seen that exchangeability is important in understanding how probabilities are assigned or deduced (Chap. 4) for propositions with finite content. And we saw above how parameters can arise by taking finite information to the limit in a well defined manner. We can get to parameters another way, using exchangeability and De Finetti's representation theory.

Suppose evidence E is such that $\Pr(Y|E) = p$, where p is non-extreme. Then if Y is exchangeable and part of an infinite sequence it can be shown that

$$\Pr(Y_1 Y_2 \ldots Y_n | E) = \int_0^1 \prod_{i=1}^n p^{Y_i}(1-p)^{1-Y_i} dQ(p). \qquad (8.13)$$

This is an existence proof, not a constructive one as above. All it says is that *some* distribution $dQ(p)$ exists, we know not what. Bayesians have formed the habit of parameterizing this "prior" distribution on p in an *ad hoc* way, which is natural since there is no guidance in (8.13). A typical choice here is to set $dQ(p) = dp$, i.e. a "flat" prior, which gives the same results as above. In a way.

It's only "in a way" because the theorem isn't needed if, as we claim, $\Pr(Y|E) = p$, because, of course, we know p and we can prove exchangeability. Everything we need to know we have deduced from E, as we did above. We only move to infinity as approximation if it helps finite calculations. Bayesians use the theorem when they assume p is unknown but—and additional assumption—that it is fixed number.

8.7 Mystery of Parameters

As long as we start with parameterized models without understanding the nature of their origin, we're courting over-certainty. Nearly all models—I speak of those in actual practice and not only those pursued by specialists—are *ad hoc*. They are used because of custom or because of ignorance of alternatives. The parameters inside the models just appear.

There are, however, natural places for parameters. For instance, Planck's "constant" h, which has the units of a physical action and is commonly expressed in Joule-seconds. This parameter is necessary to a host of equations and is part of several theories (or one grand one) of quantum mechanics. Werner Heisenberg's uncertainty relation between a particle's location and momentum, for example, is $\Delta x \Delta p = h/4\pi$. This h is *posited*; it is a premise of these theories (or of one grand theory); its value is not deduced. If it were deduced, we would know its value. We do not even know if it is *deducible*; which means we do not know if it is an epistemological helper, as parameters usually are, or it is an ontological part of nature. Is it part of us or the world? If us, it means the theories we have about motion are incomplete. If part of the world, it means we have learned something fundamental about reality.

The goal of all physical theories, as we discussed in models of causality, is to deduce from indubitable axioms *the* causes of change; all of them (some claim that extensions of string theory are just that, e.g. [214]). We are obviously not at that point, and it is not at all clear we ever will be. If we understood all, we would have a theorem, possibly monstrous in length, that said, given X, $h = this$. Because we do not have that, and we do not even know if we can have that, this is proof we do not fully understand all causes. We do have parameterized models, which means the "laws" of physics are at least partly, at this point in time, epistemological. We do not even have *proof* that "constants" are constant, i.e. that they take the same values in all parts of the universe at all times. Instead we have the assumption (our premise) of constancy (some parameters in some theories are said to change, a subject beyond this book). Again, this implies laws which contain uncertain premises are to be interpreted in an epistemological sense. It is not proof that the objects to which the laws refer are not real, only that our understanding of them is incomplete. Think of the ubiquitous coin flip: the probability which describes our uncertainty is epistemological, but the physical coins and flip mechanism are ontological, and the two things—reality and our understanding of it—are separate. Even in the coin flip we have the tacit premise of constancy; that is, that the essence or nature remain unchanged in time and space. Observation gives weight to this premise, as it does to some physical "constants", but perhaps not all of them.

As it is, h and many other similar parameters—fine-structure, speed of light, gravitational, etc.—can only be estimated through experimentation. Necessarily, there will be uncertainty associated with the measurements, which implies we cannot be certain of the value of parameters. That again shows our understanding of these parameters is at least partly epistemological. Anyway, the uncertainty inherent to the estimates is, or should be, in every situation "stuck to" the parameters—not just in physical models; all models. That means when a parameter is input into a predictive equation, the uncertainty of the parameter should be carried forward. So if we say, based on h (and other premises), that the particle will be at x (or whatever), we should instead say that the particle will be at $x\pm$ this-or-that. And this judgment holds even greater force if those other premises also contain uncertain parameters. All the uncertainty "adds up" (it may not be linear, of course), and can't be ignored unless the decisions made on the predictions are insensitive to the size of resultant uncertainty. I am not arguing against the extreme usefulness of back-of-the-envelope calculations. But I am saying that when it counts, as it often does, specification of complete uncertainty is warranted.

Now most statistical models are not used in the same way, as we shall see next chapter. Instead, an *ad hoc* parameterized model is given and the focus is put solely on the parameters *as if* the parameters were the reality. We start by wanting to know how some X is relevant to knowledge of some Y, but end with assuming falsely that some parameter Θ of X is not only X but a cause of Y. X is all but forgotten; and so, really, is Y. The Deadly Sin of Reification has struck. There were, in the past, good computational excuses for this. Calculating uncertainty of actual observables was difficult. But that is no longer so. The problem is strange because statistical model parameters are usually very unlike physical parameters, but that's because explanation and the discovery of causes is *the* goal of physics. It isn't of statistics, which has understanding of uncertainty as its destination.

Not for the last time I emphasize that problems should be set up based on the limitations of measurement and acknowledgement of finiteness. Only then can limits-as-approximations be used, *after* it is understood how these approximations work in the given problem. We need desperately to understand the origin of parameters. And we must view statistics in a predictive not parametric sense. This eliminates much over-certainty. But not all. People will constantly invent new ways of going wrong.

Chapter 9
Statistical and Physical Models

"The only useful function of a statistician is to make predictions, and thus to provide a basis for action."—W. Edwards Deming.

Statistical models are probability models and physical models are causal or deterministic or mixed causal-deterministic-probability models applied to observable propositions. It is observations which turn probability into statistics. Statistical and physical models are thus verifiable, and all use statistics in their verification. *Statistics* are summaries of or are simple observations themselves; statistics are observed propositions (or mathematical functions of them).

Classical frequentist statistical modeling emphasizes hypothesis or "significance" testing and estimation. Testing flows from the desire to know and to decide whether a proposition is true or false. And estimation comes from the wish to know how much or how strong some signal or force is. These are the correct aspirations, so it's surprising that not only is neither goal met using classical approaches, but that everybody thinks they are. Physical models usually are on the side of angels here, engaging reality often and then, as is proper, mercilessly culling models which don't match observations. Politics, however, in both statistics and physics, has been known to save unrealistic models.

Frequentist hypothesis testing and estimation are parameter-centric. And so is classical Bayesian prior-posterior analysis. These methods tell us about things *inside* models, but they are silent on what's going on outside; that is, they are mute about reality. It is all but forgotten that we can be as certain as we like about what's happening inside a model while remaining largely ignorant of what is happening in reality. To assume, as nearly everybody does, that the certainty which applies to a model's guts applies equally to reality results in gross, systematic, and, at times, ridiculous over-certainty. Entire *fields* make their living based on misunderstandings of classic statistical theory—in both its frequentist and Bayesian forms.

However careful, say, an academic statistician is and however diligently he admonishes his readers not to over-interpret results, it is true and confirmed endlessly that his students *will* over-interpret and misbehave shockingly. Why? Because

© Springer International Publishing Switzerland 2016
W. Briggs, *Uncertainty*, DOI 10.1007/978-3-319-39756-6_9

classical methods do not answer questions ordinary people ask, and nobody can remember the arcane and esoteric interpretations given to the questions it *can* answer. If you doubt this, ask any, say, sociologist for the interpretation of a hypothesis test or confidence interval. It won't even be close to correct. Or, better, look to journals which use statistics routinely. Over-certainty abounds. The blame for this situation ultimately rests on those who invent and promulgate theories, or, in other words, us. This is because the misuses of statistics are so egregious and pervasive that the reaction from the top should have long ago been abject horror, instead of the complacency we see. It's well past the time for a fundamental change in practice.

If we're doing it wrong, how do we do it right? The answer is obvious: answer questions people ask. Look outside our (mostly *ad hoc* and in many cases dubious) models and speak of reality. We must come to a proper understanding of causality in probability models. And, as will become clear, we must guard against the we-must-do-something fallacy, a disease which largely affects academics anxious to produce research. In this and the next chapter I'll often use the shorthand "statistical model", which applies to any kind of model of observables, since statistics are the main way to judge them.

9.1 The Idea

This section is addressed primarily to those who use probability and statistics but who had no part in the development of its methods; developers can skip ahead to the next section.

The idea is this: (1) Look, (2) Don't model. Very many times, simple summaries and plots of data are superior to models. But if you're going to model, follow Deming's advice: make predictions. Predictions are verifiable.

Misunderstanding statistics is causing much harm. The failings of the current methods of practicing statistics are known but largely unheeded, e.g. [9, 230]. Arguments by critics are now so common that I do not reproduce any but the most relevant here. Significance chasing, a.k.a. wee p-value hunting, is parameter- and hypothesis-centric, which inverses the normal order of scientific questioning and usually involves unobservable entities. The solution to the problem is not a simple replacement of frequentist with Bayesian prior-posterior distribution analysis. Both classical frequentist and Bayesian methods do not allow the assessment of model performance and usefulness. Both answer questions nobody but mathematicians ask, questions which are almost always irrelevant for real decisions.

What are relevant questions? You go to the doctor and he recommends a new pill. What do you want to know? The sensible thing: what are the chances this pill cures your disease. Statistics as currently designed stubbornly won't answer that question. It's worse than it sounds, because the doctor is basing his recommendation on studies which, for instance, compared this new pill against an old one, studies which pronounced whether observed differences in the pills' effects were "significant".

What did the researcher designing the study on which the doctor relied want to know? The sensible thing: what are the chances more people get better taking the new pill rather than the old. Again, statistics won't answer that. Statistics will only announce whether the results were "significant". Suppose they were. What happens next? Everybody wrongly falsely incorrectly mistakenly and without warrant assumes that significance is equivalent to knowledge that the new pill is better than the old, or that all the differences in the observations were *caused* by the new pill. "Significance" is the criterion of evidence the FDA, EPA, and every other rule-making body uses. This is a dismal state of affairs.

Here is how statistical modeling *should*, but does not, work: Compile evidence probative of some proposition of interest, and then calculate the probability this proposition is true given this evidence.

That's all of statistics in twenty words. How is this done? Simplest example in the world. Proposition of interest, "This coin comes up head in a flip." What evidence is probative? Well, the coin is two-sided, one side labeled heads the other tails, which when flipped (in such-and-such a manner) must show one of these two. The probability the proposition is true given this evidence is 1/2, as expected. But we're not restricted to this evidence. A physicist might step up and measure the coin's spin, the force of the flip, and so on. Using *that* evidence, he can come to a different probability the coin lands heads. Indeed, and as has been done, with precise enough measurements, the physicist can predict with something close to certainty what will happen.

Or we might have a two-valued "coin" or coin-like object the physics of which are unknown or where the knowledge of its properties is limited. We might compile the results from a sequence of "experiments", the causal nature of which is known in varying degrees. From this, and from the starting knowledge that we have a two-valued "process", like the coin flip we can deduce the probability the process is in one of its states in this next experiment or in the next n experiments or whatever. We can form any observable proposition we like about to-be observed process. And then we can check whether our model "works" in the sense of giving useful probabilities.

In coin flips, the difference between, say, a referee on the field and a physicist is that they have different information; they have different *models* of the situation, as it were—as it *is*. Different models give different probabilities. This is not a profound statement, yet all evidence suggests remembering it is monumentally difficult.

The trial for the new pill—or for *any* other situation—works in the same way. Gather evidence relevant to the question at hand, this evidence becomes a model from which probabilities are calculated. Now this might happen in mathematically complicated ways, but that is nothing. We don't need to understand the math to comprehend the answers. "The probability of this with respect to that" makes sense to everybody. This is statistics-as-argument. It is a predictive approach.

This form of statistics is actually used, and in more places than you might have guessed. Every time you drive through an automated toll booth the machine that takes a picture of your license plate must take the evidence it has—the picture itself, the characteristics of the kind of images stored, and the like—and calculate the probability the license is "YAC 893" or whatever. It must make a classification (a decision) based on a calculated probability and an understanding of the consequences of the decisions. If you've ever made a bet, say on the stock market or a sports team, you've used this form of statistics. Here you're painfully aware that if only you had had better evidence (a better model), you would have formed superior probabilities. And here you understand there is a difference from the probability the team will win and the decision or bet you make.

This reality-based statistics is sometimes called *predictive statistics* because it predicts what will happen (but this term has multiple meanings, so be careful); for some examples, see [88, 89, 127, 128, 144, 200]; there are parallels with so-called objective Bayes, see e.g. [225]. In this scheme, models are checked against the world. Bad models are rejected, good ones cherished. This sounds like science; or how science used to be. Science uses predictive statistics in many fields, usually the closer these are to engineering, but it is also, or used to be primarily, found in physics. The idea is to make testable predictions so that model goodness can be assessed. This is impossible in hypothesis testing or Bayesian parameter prior-posterior analysis.

9.2 The Best Model

The best model happens to be the least used among professionals, but the most resorted to by civilians, who thus gain a decided edge. The best model is no model. The best model is to just look at the data, evidence, and premises gathered and ponder them. *No* model is ever needed to tell us what we observed, the first step in gaining an understanding of essence and nature. We know what we observed because we have observed what we have observed. That sounds a useless sentence, but it isn't. It is emphasized because it is everywhere (in professional circles) doubted and perhaps even disbelieved. I have often put it to statisticians and the most positive response I have received was a blank stare.

You're a doctor (your mother is proud) and have invented a new pill, profitizol, said to cure the screaming willies. You give this pill to 100 volunteer sufferers, and to another 100 you give an identical looking placebo. Here are the facts, doc: 71 folks in the profitizol group got better, whereas only 60 in the placebo group did. Here is the question: in what group were there a greater proportion of recoverers?

Every statistician hearing that believes there's a trick, and to solve it he will propose a model, say, a "z test", or whatever. Yet the untrained—he must be untrained—civilian will say, "The drug group", which is the right answer. Of course it is!

Question two: what caused the difference in observed recovery rates? I don't know. But I do know that some thing or things caused each person in each group to get better or not. I also know that "chance" or "randomness" weren't the causes. They can't be, because they are measures of ignorance and not physical objects, as we have already seen. Lack of an allele of a certain gene can, say, cause non-recovery, or a diet of carrots in sufficient quantity can speed the recovery, but "chance" is without any power whatsoever. Results are never due to chance, they are due to real causes, which we may or may not know. If our goal is only to make the statement which group got better at greater rates, no model is needed. Why substitute perfectly good reality with a model? That is to commit the Deadly Sin of Reification, which is explored in greater detail next chapter.

If our goal is to say something about patients *not yet seen*, then a model is not only just the thing, it is required. How to build such a creature we next discover. It is without question a model is needed because we did not measure anything that could have caused the observed difference, and it is a *certainty* that more than one cause was at work. If there was only cause at work every result would always be the same, as we learned earlier; or if there were only two causes at play, then everybody would have the same outcome in each group, but each group would be different. It is also unlikely that our model will discover the pertinent causes; the best we'll be able to do is to characterize the uncertainty we have in new observations, which is usually an adequate goal.

Perhaps my admonition now seems needlessly strong. After all, it *is* obvious which group had the higher proportion of recoverers, and, the objection will continue, the experiment was run for the purposes of making statements about yet-to-observed patients, so a model was required after all. There are two reasons why, if anything, my caution was not strong enough. One is due to the various uses to which data is put, which I answer next. And the second is the "modeling reflex".

This reflex is so strong that it isn't realized that there is no model deducible from the stated premises. All we know is that pills were given to 200 folks, and this many got better and this many didn't. We know nothing about the pills or the people, not any of their demographic or biological conditions nor how many folks in the future will be eligible for the pills and so forth. There is surely positive information in our data (premises) that the drug is possibly doing something (but only sometimes) the placebo is not. But it's unlikely we'll be able to make sufficiently detailed measurements if we were to repeat the experiment such that we identify the precise causes of a cure in each individual. In order to model using the information provided, something additional has to be assumed. Perhaps this something will be right or approximately right, or again perhaps it will be wrong. There is a good case that in medical or other situations where measurement is careful, assumptions necessary to make models are often reasonable. But that case cannot be made where measurements are sloppy or crudely conducted and where theories of causation are fanciful, as they often are in some fields (those which use questionnaires, mainly, as we shall see). Models have to be justified. They usually are not. That they were done by "competent" researchers is usually the only justification offered, and that is insufficient. All probability is conditional and so all models are conditional on the premises assumed. Different premises lead to different models. The public (including politicians) are far too accepting of the justification of models used to rule and regulate their lives, and they are blissfully unaware that models are not unique. As we'll see below, there are good ways to test the assumptions that create models.

Time series data, or rather observations that occur in time, are abused the most. One example will suffice, though I'll change the names and details to protect the guilty. One author claimed that violent deaths were decreasing in time, and he offered a model and a reason why this was so. A second author claimed the first author was wrong: deaths were not decreasing, but he offered no reason beyond a model why this was so. The same, or similar data, produced two models with diametrically different conclusions. The reader will recognize many similar situations.

Both parties pointed to their models to say, "Deaths (are) (are not) decreasing". It was the model in each case which was the arbiter of the truth. This is the Deadly Sin of Reification. No model is needed to say what happened. Either author had merely to look at the data and conclude, with perfect certainty, whether deaths were increasing, decreasing, or holding steady. Only one small additional premise is needed: a definition of "decreasing" (or its obverse "increasing"). As stated at the beginning of this book, the meaning of the words used in argument are themselves tacit premises. These tacit premise must be made to "come out into the open" when arguing formally. Next chapter, we'll see the mistakes which are made when the tacit premises are left tacit when I discuss "trends" in time series. For now, only assume that a definition is had and is agreed to by all parties: "decrease" means this, "increase" that, and "hold steady" something else. I beg the reader will attend closely: with the definition settled, all we have to do is to *look*. Either the conditions of the definition will have been met, or they will not have been. And since, it is presumed, these three categories are exhaustive (about death trends), somebody will be certainly right and the other certainly wrong. There is no need to argue! And with even greater force, there is no need, there is absolutely no need, for any kind of model.

Models in these cases are *replacements* for reality. What possible reason could there be to replace a perfectly good reality with a fiction? I ask because it is done all the time. Each was sure of his model and (of course) his cause. That's what really caused the dispute. Love of models. Scientists these days are like modern-day Pygmalions, falling in love with their creations. Only they are not as blessed as Pygmalion; a scientist's model is forever lifeless.

Now if it were true that the data in a given situation were measured with error and we wanted to quantify the uncertainty in what the truth might have been, then we need a model, just as we need a model if we want to quantify the uncertainty in what the future will be. Models are used to quantify the uncertainty in what we don't know; when we are certain, they are not needed. Unknown is unknown. On the other hand, if the data is known *it is known*. Don't play with it. Unnecessary fiddling is rife. As I say next chapter, an entire book could be written of the abuses spoken of in this small section, but that will have to wait for another day.

But aren't models also needed for hypothesis tests, to see if the observed differences are "due to chance"? No. This is always a mistake: hypothesis testing should never be used. This is proved below.

9.3   Second-Best Models

The goal of modeling should be this: to gather evidence X to make probability statements about propositions Y. Schematically, as in the last chapter (this is expanded in a practical sense several sections below):

$$\Pr(Y|X). \tag{9.1}$$

Although probability is general and places no restrictions on the nature of the propositions Y, statistics and the physical sciences by custom and as a useful division of labor, but not of philosophy, restricts Y, and much of X, to be observable (contingent). X will contain past observations and other information, including deductions and possibly causal or logical propositions, thought probative of Y. *Premise, assumption,* or *observation* as names for X are preferred over *variable*, because premise and the like are easily seen to be what they are: assumptions. "Variable", unless one is careful, can led to reification and mistakes in assigning cause. There are no parameters here, nor are there decisions: this model is pure probability. In this section, only the idea behind this approach is given; there are some suggestions about how to go about this at the end of this chapter, and specific implementations in the next chapter.

Most unfortunately, statistics as practiced is rarely like (9.1) and is something else entirely. Which is to say, the material that follows is *not* standard—but should be. Statistics as classically practiced is like this:

$$Y \sim D(X, \theta), \tag{9.2}$$

where the observable Y is said to be "distributed" according to some probability distribution D, which is a function of premises (usually a smaller set than in (9.1)) X, themselves indexed by parameters θ. Now (9.2) can be turned into (9.1) by "integrating out" the uncertainty in the unobservable parameters. The operation is an integration over (9.2) if θ is continuous, else it is a sum. In other words, given X, the probability of Y is the value of (9.2) weighted by the uncertainty in θ. Schematically:

$$\Pr(Y|X) = \sum_i \Pr(Y|X, \theta_i) \Pr(\theta = \theta_i|X). \tag{9.3}$$

where X "contains" everything we know, including past observations of the proposition of interest and probative observational data, other premises specifying the model, and specifications of new values of the probative data. (This is a cartoon equation, which is made specific in individual implementations.)

In modeling as commonly practiced, discussion settles on measures and statements about the parameters, about, that is, objects like $\Pr(X|\theta = h)$ in frequentist statistics or like $\Pr(\theta = h|X)$ in Bayesian, for some value of the parameter h

(this can be a set). The remainder of the equation is forgotten. Because all or most of the math (as is proper) is set aside when communicating results, classical methods make errors in inference hard to spot, and it has led to a ritualized form of statistics. Decisions about Y based on statements about parameters mixes up probability and decision. Even in "non-parametric" statistics, the goal is decision not probability.

In the predictive approach, as given in (9.1) or (9.3), measures of relevance should replace hypothesis testing, and direct calculation of propositions of actual interest should replace estimation. The classical methods of testing and estimation are discussed below. The replacement in both cases is the same; which is to say, Eq. (9.1) or (9.3) and not (9.2). The onus of decision should be removed from the method and put where it belongs, on the narrow shoulders of users. Statistics must become less like ritual and more like hard work. Statistical pronouncements must be put in the form where they can be verified—independently verified by direct comparison with reality. Models which fail to conform to reality are to be expunged.

Some rough but familiar examples first. Suppose X is a compound proposition about past observations, nature of employment, residence and the like, and sex (M or F), and Y is the proposition Y = "Income is greater than fifty thousand." We might be interested in (with obvious shorthand notation) $\Pr(Y|X_M)$ and $\Pr(Y|X_F)$. If both of these are equal, then knowledge of sex is irrelevant to knowledge of Y—given the remainder of what's in X. If the other constituents of X change, sex might become relevant to Y (as was shown previously). If Y is modified to a different obtainable monetary figure, say $Y' = $ "Income is greater than seventy thousand", and $\Pr(Y'|X_M) = \Pr(Y'|X_F)$, then again, knowledge of sex is irrelevant to knowledge of (as is obvious) Income at this level—assuming again, of course, *the other conditions in X apply.* Of course, there may be amounts of Y at which sex is relevant. Plots can be made in which Y indexed by amounts forms the abscissa and the probabilities calculated assuming the X of interest the ordinate (reversing the x-y!). In other words, plot y by the probability of Y = "Income is greater than y" by conditioned on whatever (combination of) X is of interest.

If, say, $\Pr(Y|X_M) = \Pr(Y|X_F) + \epsilon$ for some small ϵ, then knowledge of sex is *relevant* to knowledge of Y (and its stated income figure; recall Y is a fixed proposition). Whether ϵ is "large" and "important" or "small" and "negligible" or whatever are *not statistical questions. Any $\epsilon > 0$ is enough to prove relevance. Whether this is "practical" relevance or not depends on the decisions to be made with the information. The size of ϵ might be important in one context and ignorable in another. Suppose $\Pr(Y|X_M) = 0.4$ and $\Pr(Y|X_F) = 0.45$. There is a higher chance that women, given what we know in X, will make the income stated in Y. Is that extra 5 % "enough" to make a difference? That depends on what decisions are going to be made of these probabilities. Is somebody going to be sued? How many people matching the premises implied by X exist? Probability is silent—*and should be*—on the import of any number. Now it might be that the person making the calculation may judge this difference of 0.05 probability negligible. That being so, sex can be removed from X, i.e. stricken from the model. In more classical language sex is removed from the model, making the model more parsimonious. Or it might be kept in and used for downstream decisions.

To emphasize: there is no telling what ϵ is important. None. An $\epsilon = 0.01$ may be crucial to one man and less than trivial to another. Relevance is *always* an extra-statistical question. Its importance is *always* conditional on outside criteria. May the first man who says, "Why not make $\epsilon = 0.05$ be the standard?" be anathema.

If an element, such as sex in the example, was among (or was) the main reasons for a study, then it can be argued that no $\epsilon > 0$ is too large to exclude that element from the model. It is the value of ϵ that should be reported to the world. Let each make of it what they will. Of course, the investigator can make of it as he wills, too. He can show the consequences of its removal or of retaining the element under circumstances he judges interesting. That is, as it sounds, a lot of work; vastly more effort than throwing data at a software package and hunting for wee *p*-values. But this approach is vastly more honest.

Another example. Suppose Y = "This person has COPD" and X is a compound proposition about the nature of a particular group of people including their body mass index (BMI), a weak and somewhat inaccurate measure of obesity. If $\Pr(Y|X_{BMI=29}) = \Pr(Y|X_{BMI=30})$ then knowledge of differences of BMI at these two levels is irrelevant to knowledge of Y. A plot may be made of $\Pr(Y|X_{BMI=x})$ by x. If this probability varies at all, BMI is relevant to Y, given the remainder of X (each premise in X will be fixed at some value). If the probability does not vary, BMI is irrelevant, given X. If the levels do vary, what is important not only depends on the difference between BMI levels, but what BMI's are "actionable". This depends both on measurement and on the consequences of decisions. We discuss measurement next. What levels of BMIs are "expected" (where I use this word in its plain-English sense)? That could depend on premises partly in X, and partly not in X. That is, outside information may have to be incorporated to see how serious or how likely any differences in probabilities between different levels of BMI would be.

Another example. A group of persons with measured (medical) characteristics X. We're interested in Y = "Person p will live past time t" given his characteristics and the information from the measured people. Note very carefully that we're talking about *future* events for this person: if he is already dead, we know that and don't need probability. Compute $\Pr(Y|X_p)$, where X_p represents the characteristics of the measured people *and* person p. A plot of t by this probability can be made. This is, of course, survival analysis, but it differs in that this curve will not have uncertainty in it: there will be no "confidence" or "credible" bounds. The probability is the direct prediction for *this* person who has certain stated characteristics. Plots for fictional or representative persons having stated characteristics can be made in the obvious way. Relevance is ascertained as before: if the survival curves do not differ for different levels or values of some pertinent characteristic, then this characteristic is irrelevant (given the others) else it is relevant.

Another example. A group of objects belong each to one of several categories. Past data on similar objects and their category membership is available; all this is X. We want Y = "This object belongs to category c". Compute $\Pr(Y|X_c)$. This is classification. Relevance is the same as before. Again, this is a direct prediction with no extra bounds or uncertainty. Further, it is stated as a direct probability and easily interpreted.

Keen observers will have noticed that there are no uncertainty bounds, "confidence" or "credible" intervals and the like, not just in the survival analysis example, but nowhere. They are not there because they are not needed. Why? Because no parameters exist in (9.1). Parameters might appear in the math which facilitates calculations, but they are "integrated" out at the end, as they are in (9.3). We are making predictions, stating strengths of associations; we are not producing statements about unobservable and uninteresting parameters.

9.4   Relevance and Importance

Recalling Chap. 8, *in-model* relevance is when

$$\Pr(Y|XW_i) \neq \Pr(Y|XW_j), \; i \neq j, \tag{9.4}$$

where X are the "main" premises and W the premises under test, and where W can take more than one level or value. All that is required for relevance is an inequality of probabilities for at least two different levels or values of W. Any inequality, not matter how small, produces relevance. Premises which do not exhibit any in-model relevance—there is equality for each i, j—are irrelevant, which is better phrased as *out-model* irrelevance. This is when

$$\Pr(Y|XW) = \Pr(Y|X). \tag{9.5}$$

Adding W has done nothing, in the presence of X, to add to our understanding of Y. It is not necessary to the model. Any premise deducible from X is irrelevant. Any necessary truth, such as a tautology, is irrelevant. In particular, as we saw earlier, the tautology "Y can happen or it can't" is irrelevant for understanding Y. Again, any inequality, no matter how small, *proves* relevance.

Equation (9.4) is ripe for plotting, but with a strong proviso. The values of W_j take the x-axis and the probabilities in (9.4) the y-axis. Departures from irrelevance are easily spotted. The proviso is: for a fixed X. Which X to pick? This question stresses the difference between the predictive and the classical approach. Those X which are felt important to a decision maker are the X that should appear in (9.4). It could be that W is irrelevant or judged unimportant for those X which are of no decisionable interest, but W could still be relevant for other X. The decision about which X is *not* statistical. If an author is trying to convince an audience of the importance of some W, it is up to him to pick those X which are convincing to that audience. This also allows the audience, as the predictive approach insists, to verify the model independently of the author. This is the basis of true and not interested replication.

For understanding how the model works with the data at hand, obvious candidates for (9.4) are those X we have already seen. That is, each "data point" is used to compute

$$p_i = \Pr(Y_i|X_i). \tag{9.6}$$

And then p_i may be plotted (or otherwise examined) for each premise in X. For example, suppose X = "X_a X_b X_c" (the conjunction). The plots (or examinations) of $X_{a,i}$ etc. by p_i may be made. If Y is multi-valued, some creativity is needed to display all probabilities at once, but it can be done.

The same sorts of assessments may be made for the addition or subtraction of data points. Adding more data usually adds information, but it doesn't have to. For a fixed X (or fixed XW conjunction), how much does adding new data points change our understanding of Y? That can be plotted. This is done *predictively*. We imagine, or assume, the new X. That means we have to specify the values of all X. Again, which to vary and which to hold fix depends on the decisions to be made. A cruder approach would be to see how each new data point changes a summary measure of information, such as entropy. This leads us to importance.

Importance—probabilistic and *not* practical importance—is the level of influence a premise (variable) has in the model, given the other premises (variables). It is therefore related to the probability Y is true given that premise and given the others. A premise which cannot take "levels" can only exhibit *out-model* importance. Either it is in the list of premises or it isn't. The corresponding change in the probability of Y is its importance. But some premises are variable; e.g., sex. The changing probabilities of Y as the premise steps through its levels is the importance of that premise. There will arise a clamor for single-number summaries of importance. This should be resisted, and fought off by armed force if need be. It won't be, though. So here are some possibilities.

Out-model importance is easy. Compute $p = \Pr(Y|XW)$ and $q = \Pr(Y|X)$, just as we would for testing relevance. The absolute difference $|p-q|$ is the importance. If W is irrelevant $|p - q| = 0$. This is obviously bounded in $(0, 1)$. It is also *conditional* on X (and on values y might take inside Y). If X is not multileveled or multi-valued, we're done. But most X will be. Therefore, one thing we can do is step through the data and compute $p_i = \Pr(Y_i|X_iW_i)$ and $q_i = \Pr(Y_i|X_i)$ and form the collection $p_i - q_i$. If W is irrelevant for all X then $p_i - q_i = 0, \forall i$. W may be irrelevant for some i, which is to say, some X.

What to do with the collection $p_i - q_i$? What indeed. Take its mean, compute its entropy or variance. All summaries remove information, and each kind of summary removes different parts. Entropy has additional reasons for use, as are well known (I leave aside the various methods for calculating it). Importance is related to information, and information is often *summarized*—and *not* fully specified—by entropy, see e.g. [136]. W is more statistically—and not necessarily practically— relevant the larger entropy of the collection $p_i - q_i$ is. Practical, which is to say *real-life*, relevance is assessed as given above, by specifying those premises of real interest to decisions.

These calculations are *not*, incidentally, defined with respect to the more usual statistical expected value of W given X. Instead, the suggestion it to compute the average entropy of the W at each of the levels it takes given a fixed X. In order to calculate statistically defined expected value we need to have information external to X, say Z, which is probative of W, so that we can derive probabilities of W. In particular, W cannot be deducible from X because if it were it is by definition redundant.

Guidelines for relevance "levels" are *not* recommended, for whatever summary is computed. The last thing science needs is another magic number. No one anywhere ever should recommend anything remotely like, "If the difference in probabilities (or entropies or etc.) is greater than τ, the results are 'significant.'" Significance depends on decision, and decision is not probability.

9.5   Measurement

This subject was discussed briefly last chapter in the section on parameters, which should be reviewed first. It is meant as a compliment and amplification of the idea that models of observables should start out discrete and finite, and only passed to the limit in certain, specific ways. Infinite-valued models will result as an approximation to real-life measurement. How good these approximations are can only be discovered if the finite-discrete models are known.

No measurement can be infinite in scope or refinement. Every measurement is instead discrete and finite. For example, a digital thermometer has a minimum and maximum detectable temperature and can only display at fixed marks or output. An analogue thermometer must be interpreted in some fashion but has the same kind of limitations. Income can only be measured to the nearest cent (or whatever), and infinite incomes are impossible. A physics experiment to measure reflected light must use a device which gives limited information. Computers with off-line storage can calculate large numbers, but none can calculate infinite ones, and all storage is discrete and finite. And on and on for every real thing. Now whether there be actual infinities of anything, or whether infinities are always potential and not actual, is a separate and fascinating question, just as are whether our minds can grasp the infinite, and, if so, to what extent. See [186] for a discussion. But in the sciences, we must take reality in pieces and this has consequences for our understanding of uncertainty, which is again epistemological and not ontological.

Measurement is closely related to decision. Suppose there is interest in grade point average. If grades are awarded on a numerical scale, and since only a finite number of classes can be taken and discrete numerical grades earned in each, only a discrete set of values of GPA can result. If the number of students tracked is large, the set of possible grades is also large—but always discrete and finite. Perhaps some values in this possible GPA set are 3.00119, 3.00120, 3.00121, and so forth. These numbers are different (not equal). But are they different enough such that different decisions would be made informed by probability calculations with these numbers as functions? That depends on the context. With one exception, it is difficult but not impossible to imagine real-life scenarios where these numbers are not all "equal to" 3—practically speaking. The exception is to award the best or worst GPAs. There, 3.00121 trumps 3.00120.

Measurement should thus be tied to decision. Any analysis should come after understanding what decisions will be made with the measurements. For instance, we might want to know, given suitable evidence, the probability the top student's GPA will be exceeded. If that GPA was 3.00121, the answer can surely be calculated, but, as anyone familiar with these kinds of models understands, this will require an enormous sample size. It depends on the premises/model, but the answer to that question and the question "What is the probability the GPA exceeds 3" will most likely differ by a trivial amount—where "trivial" is in the context of most real-world decisions. Thus it is better to further compact the measurements. Perhaps every hundredth or even tenth of a point is sufficient; or even to the nearest whole number for simple decisions. Collapsing saves effort, tailors analyses to decisions, and makes for cleaner reporting, which in turn tempers over-certainty. How many times have you read papers which said something like the mean age for some group under study was 38.145? That "0.145" after the 38 corresponds to just over 52.8 days; and if this fractional part were decremented to 0.144, it would correspond to just under 52.5 days. Are our decisions really going to fall to these measly difference in days? Worse, those superfluous digits instill over-certainty, the false feeling that *science* is happening.

It will be objected that collapsing measurements loses information. This is so. But the information was by the collapse put into the form useful for the decision at hand. Collapsing removes *extraneous* information for *this* decision. It clarifies. Collapsing only loses information for decisions, and thus analyses, which require the finer yardstick. Of course, collapsing already occurs when using histograms, when summarizing responses to survey questions and on and on. There are questions about *how*, say with histograms, to pick the bin widths and edges, but nobody said it had to be easy. Putting the data into the form ready for decisions is eminently sensible. Collapsing also allowed easier use of parameter-free, deduced models (which were described in Chap. 8). Of course, collapsing does not have to be done. Measurements are still discrete and finite.

Suppose a many-objects-in-bins example, resembling a classical multinomial. Each measurable GPA gets a bin. If GPAs were measured to the nearest thousandth than to the nearest tenth, many more measurements would have to be taken to flesh out our uncertainty because unless the sample size were enormous most "bins" would likely be empty. This makes predictions problematic. Not conceptually problematic, but computationally. Collapsing measurement requires more work up front than is traditional, because it requires taking a view of how the decisions fit in with the uncertainty. But since the tradition is what is in question, this objection does hold much force. Another objection is that methods for discrete, finite, and collapsed data are not developed or are not widely available. This is very true. Regression, the most used of all techniques, does not fit into the framework developed here. Discrete regression is categorical data analysis and can certainly be done (see [2] for what I mean by categorical data analysis). I leave it to mathematically adept readers in search of problems to discover.

Traditional analysis of some quantities presume infinite values, as discussed when investigating the nature of parameters. Take income. Regression and other traditional methods assume it can increase without limit or be comprised of infinite gradations. As an eventual *approximation*, this might be fine: it depends on the decisions to be made. The approximation is likely to be worse the "chunkier" the real measurements.

There is more that can be said. Suppose an object y is known to lie at one of m locations. If a measuring device which has no error were to assess the object's location, we would know that location with certainty. The inerrancy of the measurement and the possible locations are premises; i.e., they are the model. Such a model is not uncommon nor barren. "To the nearest year, how old are you?", "How many people are in the room?", "Are you male or female?" (the latter is, of course, subject to error), and many more. If we didn't have this model, we could never take or trust most measurements.

Suppose under this model and a static reality two measurements are taken in succession. Since nothing changes, both measurements must be some y_i. But, for instance, if the first is y_2 and the second y_4, the model has been falsified. Either the object moved between measurements or there was error injected into one or both measurements, or both the object moved and the measurement erred. There is no indication of cause in this case. If measurements differ but the premise of immovability of the object y is kept, our model is of measurement error. Now if we knew at each instance the nature of the measurement error, then again we would know the location of the object with certainty. For instance, if we knew the device added error 3 units up (or whatever), then we would deduce that the object was really 3 units down. But this kind of information about the nature of the error isn't usually provided. Instead, under most assumptions of measurement error, we only know that error of some kind has been added, but where the exact cause of error is not known, though some details of the cause might be, e.g. the error is additive. We might have a premise like "the additive error may be one of these numbers $e_1, e_2, \ldots, e_p$". This premise implies that every time a measurement is taken, the error can be different and is one of the set $e_1, e_2, \ldots, e_p$, but which of these applies each measurement we do not know.

Assuming this model and adding the premises that n measures $x_i = y_j + e_i$ were taken (we see x_i but y_j is the static truth), we can form the average

$$s = n^{-1} \sum_i x_i = y_j + n^{-1} \sum_i e_i,$$

where y_j says the location of the object is y_j. Given the model, i.e. the premises that the object is at one of the m locations $y_1, \ldots, y_m$ and that the error for each measurement can be one of $e_1, \ldots, e_p$, s must take one of the values

$$y_1 + n^{-1} \sum_i e_1 = y_1 + e_1$$

or

$$y_1 + n^{-1} \sum_{i=1}^{i=n-1} e_1 + n^{-1} e_2, \ldots, y_n + n^{-1} \sum_i e_p = y_n + e_p$$

This makes it natural to think of how far off the average is, or rather, how much measurement error there is in the average.

For example, let the possible errors be $-1, 0, 1$. If we take $n = 1$ measurement, the chance the average is error-free is 1/3, which is also the chance the error is -1 and the chance it is 1. If $n = 2$ measurements are taken, the error can be in the set $-2/2, -1/2, 0/2, 1/2, 2/2$ with corresponding probabilities $1/9, 2/9, 3/9, 2/9, 1/9$. By the time we have $n = 4$, the possible errors are

$$\{-4/4, -3/4, -2/4, -1/4, 0/4, 1/4, 2/4, 3/4, 4/4\}$$

with probabilities

$$\{1/81, 4/81, 10/81, 16/81, 19/81, 16/81, 10/81, 4/81, 1/81\}.$$

There are thus two strategies. Take the measurements, form the average and then list the probabilities of all possible errors, or pick a threshold of acceptable error in the average and the probability the observed error is at least this small, and solve for the number of measurements that must be taken. If this error is, say, 0.2 and the acceptable chance is 80 %, then at least $n = 5$ measurements must be taken.

It is of interest to let the number of possible errors expand or the number of measurements to become "large", or both. Done suitably, as before, parameters will emerge, which might be useful as *approximations* to the exact answers.

9.6 Hypothesis Testing

Classical hypothesis testing is founded on the fallacy of the false dichotomy. The false dichotomy says of two hypotheses that if one hypothesis is false, the other must be true. Thus a sociologist will say, "I've decided my null is false, therefore the contrary of the null must be true." This statement is close, but it isn't the fallacy, because classical theory supports his pronouncement, but only because so-called nulls are stated in such impossible terms that nulls for nearly all problems are necessarily false, thus the contrary's of nulls are necessarily true. The sociologist is stating something like a tautology, which adds nothing to anybody's stock of knowledge. It would *be* a tautology were it not for his *decision* that the null is false, a decision which is not based upon probability.

To achieve the fallacy, and achieve it effortlessly, we have to employ (what we can call) the fallacy of misplaced causation. Our sociologist will form a null which says, "Men and women are no different with respect to this measurement." After he rejects this impossibility, as he should, he will say, "Men and women are different" with the implication being this difference is *caused* by whatever mechanism he has in mind, perhaps "sexism" or something trendy. In other words, to him, the null means the cause is not operative and the alternate means that it is. This is clearly a false dichotomy. And one which is embraced, as I said, by entire fields, and by most civilians who consume statistical results.

Now most statistical models involve continuity in their objects of interest and parameters. As before, a parameterized model is $Y \sim D(X, \theta)$ where the θ in particular is continuous (and usually a vector). The "null" will be something like $\theta_j = 0$, where one of the constituents j of θ is set equal to a constant, usually 0, which is said to be "no effect" and which everybody interprets as "no cause" of Y. Given continuity (and whatever other premises go into D) the probability $\theta_j = 0$ is 0, which means nulls are always false. Technicalities in measure theory are added about "sets of measure 0" which make no difference here. The point is, on the evidence accepted by the modeler, the nulls can't be true, thus the alternates, $\theta_j \neq 0$, are always true. Meaning the alternative of "the cause I thought of did this" is embraced.

If the alternates are always true, why aren't they always acknowledged? Because, again, decision has been conflated with probability. *P*-values, which have nothing to do with any question anybody in real life ever asks, enter the picture. A wee *p*-value allows the modeler to *decide* the alternate is true, while an unpublishable one makes him *decide* the null is true. Of course, classical theory strictly forbids "accepting", which is to say deciding, a null is true. The tortured Popperian language is "fail to reject". But the theory is like those old "SPEED LIMIT 55 MPH" signs on freeways. Everybody ignores them. Classical theory forbids stating the probability a hypothesis is true or false, a bizarre restriction. That restriction is the cause of the troubles.

Invariably, hunger for certainty of causes drives most statistical error. The false dichotomy used by researchers is an awful mistake to commit in the sense that it is easily avoided. But it isn't avoided. It is welcomed. And the reason it is welcomed is that this fallacy is a guaranteed generator of research, papers, grants, and so on. Two examples, one brief and one in nauseating detail will prove this.

Suppose a standard, out-of-the-box regression model is used to "explain" a "happiness score", with explanatory premise sex. There will be a parameter in this model tied to sex with a null that the parameter equals 0. Let this be believed. It will then be announced, quite falsely, that "there is no difference between men and women related to this happiness score", or, worse, "men and women are equally happy." The latter error compounds the statistical mistake with the preposterous belief that some score can perfectly measure happiness—when all that happened was that a group of people filled out some arbitrary survey. And unless the survey, for instance, were of only one man and one woman, and the possible faux-quantified scores few in number so that a tie is likely, then it is extremely unlikely that men and women in the sample scored equally.

Again, statistics can say nothing about *why* men and women would score differently or the same. Yet hypothesis testing always loosely implies causes were discovered or dismissed. We should be limited to statements like, "Given the natures of the survey and of the folks questioned, the probability another man scores higher than another woman is 55%" (or whatever number). That 55 % may be ignorable or again it may be of great interest. It depends on the uses to which the model are put. Further, statements like these do not as strongly imply that it was some fundamental difference between the sexes that caused the answer. Though given our past experience with statistics, it is likely many will still fixate on the causal possibility. Why isn't sex a cause here? Well, it may have been some difference besides sex in the two groups was the cause *or causes*. Say the men were all surveyed coming out of a bar and the women a mall. Who knows? *We* don't. Not if all we are told are the results.

It is the same story if the null is "rejected". No cause is certain or implied. Yet everyone takes the rejection as proof positive that causation has been dismissed. And this is true, in its way. Some thing or things still *caused* the observed scores. It's only that the cause might not have been related to sex.

If the null were accepted we might still say "Given the natures of the survey and of the folks questioned, the probability another man scores higher than another woman is 55%". And it could be, after gathering a larger sample, we reject the null but that the difference probability is now 51 %. The hypothesis moves from lesser to greater certainty, while the realistic probability moves from greater to lesser. This often occurs, particularly in regressions. Variables which were statistically "significant" barely cause the realistic probability needle to nudge, whereas "non-significant" variables can make it swing wildly. That is because hypothesis testing often misleads. This is also well known, for instance in medicine under the name "clinical" versus statistical "significance."

It may be—and this is a situation not in the least unusual—that the series of "happiness" questions are *ad hoc* and subject to much dispute, and that the people filling out the survey are a bunch of bored college kids hoping to boost their grades, see [111] on WEIRD people, who form the backbone of many studies, where WEIRD = "Western, Educated, Industrialized, Rich, and Democratic". Then if the result is "Given the natures of the survey and of the folks questioned, the probability another man scores higher than another woman is 50.03%", the researcher would have to say "I couldn't tell much about the difference between men and women in this situation." This is an admission of failure. The researcher was *hoping* to find a difference. He did, but it is almost surely trivial. How much better for his career would it be if instead he could say, "Men and women were different, $p < 0.001$"? A wee p then provides the freedom to speculate about what *caused* this difference.

Finally, here is an extended analysis showing how hideously difficult it can be to discover a cause in the simplest of situations, using an example known to many historians of statistics. A certain English lady claimed to be able to tell whether her tea or milk was poured first into her cup. The statistician and geneticist Ronald Fisher put her to the test by presenting her four cups with the tea poured first and four cups with the milk poured first. The lady did not know the order of the cups.

Can we use this experiment to discover whether the lady has the ability she claims? We only have the evidence of *this one test*. The situation seems straightforward enough, but there are hidden depths. The difficulty lies in defining *has the ability*. We cannot afford to be sloppy here.

Which of these best describes "has the ability":

- She always guesses correctly (she is never wrong);
- In any experiment with N cups, she always gets at least $N/2$ right;
- In any experiment with N cups, she might get at least $N/2$ right;
- She always guesses correctly when the tea is poured first, but will sometimes guess wrongly when the milk is poured first;
- She always guesses correctly when the milk is poured first, but will sometimes guess wrongly when the tea is poured first;
- She guesses all cups correctly until the Mth cup ($M < N$), after which her palate becomes fatigued. M may depend upon a host of factors, such as the time of day, the food she at earlier that morning, her mental attitude, and so forth;
- She guesses at least $M/2$ cups correctly until the Mth cup ($M < N$), after which her palate becomes numb, etc.?

I could have expanded this list easily. For example, "She always guesses at least $W/2$ cups correctly when the tea is poured first, but will sometimes guess wrongly when the milk is poured first, where she is presented with $2W = N$ total cups." Some of these lead to tricky counting, because if, say, she always guesses the tea-first cups correctly, *and* these come first in the sequence, *and* she assumes she knows these guesses are correct, after she sees $N/2$ cups she *knows* all the rest will be milk-cup first and she will therefore guess accordingly.

None of these definitions is in any way strange: each could really be what we mean when we say this lady knows her elevens. Where is the classical "She guesses better than chance?" Are you sure it's not already there? The phrase *guesses better than chance* must be an idiom, because as we have learned, *chance* is not causative; that is, *chance* cannot be presented with cups of tea and asked to guess. So what is it idiomatic for?

Imagine an experiment where you are presented with N cups, but you do not touch, sniff, taste, or see inside these cup. You do not even see or know who places them in front of you; indeed, the cups can be left in a distant room, miles away from you. However, you must still make a guess whether the tea or milk was poured first into these occult cups. You could guess none right, or just 1, or just 2, and so on up to all N. What is the probability that you guess none right? Because our evidence (or premises) do not specify any known causal path for you to guess correctly, and because there is a natural ordering of guesses, we *deduce* via the statistical syllogism the probability you guess any individual cup correctly equals 1/2. As long as you are not told whether your prior guesses were correct, this probability remains fixed. This lack of feedback success becomes extremely important in, say, ESP experiments. If the subject knows how many successes and failures he has had, and the total number of guesses, he could use this information like in card counting to modify his future guesses. An example of how this plays is given below. See [57] for mathematical details.

Here, you are not asked to guess the *sequence*, but whether tea or milk was poured first; i.e. we want to know the number of your correct guesses and are not interested in the order of these guesses. Also notice that there is no information in these premises that suppose there will be an equal number of tea-first and milk-first cups. But even if there were, even if we knew there were equal numbers of each and thus that there were 2^N possible sequences of cups, we are still not interested in the probability of your particular *guessing sequence*, but only in the total correct.

The uncertainty in the number you guess correctly—given no causal path—thus follows a binomial (if we don't know how many of each cup; if we do, it's something else). Importantly, you could guess, and we could figure the probability of your guessing, none right, or just 1, or 2, or even all. So, "guessing by chance" must mean the ability to guess *any* number correctly. Since you can and will guess some number (even all) correctly, you cannot "guess better than chance." There is circularity. No matter if you get 0 right, 1 right, up to N right, all are consistent with guessing by chance. But we have at least learned that "by chance" means "by no (known) causal path."

Now suppose it's you against the lady; same lack of causal path for you, and her using all her powers. Who will win? If she always guesses correctly, then at best you could only match her. The probability of matching is $(1/2)^N$, which makes the probability of her beating you $1 - (1/2)^N$. We deduce this assuming she never fails. Similarly, if we assume that "had the ability" means that "in any experiment with N cups, she always gets at least $N/2$ right", and although the math is slightly more complicated, we could also calculate the probability of you tying, losing to her, or even winning.

We could go through each of our definitions of "has the ability" (and more like them) and calculate probabilities of you winning, losing, or tying. But none of these exercises tells us which of these definitions is true, or which is more likely true than another. For that, we must turn our thinking around.

We want to know whether for this sweet old lady "has the ability" is true or false, or if not true or false, then with what probability it might be true. To judge this probability we have the evidence of our experimental setup, and whatever facts may be deduced from these premises. We also have the evidence of the experiment itself: how many cups she got right and wrong. Can we agree that we should only use this information and no other? I mean, we should only use the evidence of what happened. What didn't happen and what we cannot deduce from our experimental setup is information which is entirely irrelevant. So for example if we gave the lady $N = 8$ cups, it is irrelevant that we could have given her $N = 50$ cups, or whatever. We gave her 8 and we have to deal with just that information. We do not want to fool or distract ourselves. These are of course is trivial requirements, but I put them there to focus the mind on *the* question.

Now, if we accept that "has the ability" means "She always guesses correctly", then the probability that the lady correctly identifies any cup placed before her must be 1. This phrase is also our model. I mean, "She always guesses correctly" is our model, our theory, our hypothesis.

Why did we assume this particular model? The choice was up to us. It is one interpretation of—it naturally follows from—"has the ability." Given this model/hypothesis, and *before* putting her to the test, what is the probability distribution for our uncertainty in her guessing correctly none right, just 1 right, just 2 right, etc., up to all N right? It is 0 for all numbers except for N, where it is 1. But suppose we run our experiment and she correctly identifies only $3 < N$ cups. Given just our model, what is the probability that she guesses 3 correct? Again, 0. This proves the principle that any (logical) argument can only be judged by the premises given, and by no other information.

However, suppose we conjoin our model with our observation "She always guess correctly *and* She guessed $3 < N$ correctly" and, conditioning on this joint statement, re-ask what is the probability that she guess 3 correct? It is unanswerable because we are conditioning on a contradiction, a statement which is necessarily false. Actually, given this necessary falsity, we could derive any numerical value for guessing 3 correct, but this is obviously absurd.

We have two probabilities, the first of which is:

$$\Pr(\text{Guesses } 3 < N \text{ correctly} | \text{Always guess correctly}) = 0.$$

But we can turn the question around and ask

$$\Pr(\text{Always guess correctly} | \text{Guesses } 3 < N \text{ correctly}),$$

which is obviously 0 (and understanding there is additional evidence about the experimental set up in the probabilities but suppressed here in notation). This is a rare instance where we have falsified a model—a situation only possibly when a model says "Y *cannot* be" yet Y obtains or occurs. That *cannot* is dogmatic: it means just what it says, X is impossible—not unlikely—but *impossible.*

Now, *the* question is this:

$$\Pr(\text{Has the ability} | \text{Guessed } M \text{ out of } N, \text{Experiment premises}), \qquad (9.7)$$

where "has the ability" is for us to define (such as "always guesses correctly"), M and N are observations of the experiment, where we also take care to consider the Experimental set up (from this we know what N is, etc.).

Asking (9.7) the probability a model is true is a natural question in Bayesian probability, but not in frequentism where any statement/question must be embedded in an infinite sequence of "similar, but randomly different" statements/questions. It is difficult, perhaps impossible, to discover in what unique infinite sequence this (or any) model-statement lies. I hope you understand how limiting this is. Of course, it is possible to develop non-theory-dependent rules-of-thumb for deciding a model's truth or falsity, but any true theory of probability must be able to answer any question put to it in a non-*ad hoc* manner.

What about the rest of our models/interpretations of "has the ability"? We last time outlined several possibilities, each of them consonant with the phrase "has the ability." Which of these is the correct model and which are incorrect? That is *up to us*. It is an extra-logical, extra-probability question—at least with respect the premises we have allowed ourselves in this experiment.

Now, we could go through a similar procedure as above and calculate the probability each interpretation is true. That is, if we do not have a fixed idea in advance which interpretation (model) is true, we could use the evidence from the experiment to tell us which is most likely than any of the others. However, we must start from somewhere: some external evidence must tell us how likely each of these models is before we begin the experiment. It doesn't matter what this external evidence is; it merely must exist. The most common evidence allows us to derive that each is equally likely (before the experiment commences), but that's rather arbitrary.

Let us now assume a definite model structure and see where it gets us. We suppose the lady guesses each cup correctly or not, that she knows she will see an equal number of tea-first and milk-first cups, and that she is provided no feedback about the correctness of her guesses; we assume her palate never fatigues and that her "hit rate" is the same for either cup type. We will not assume perfection, but we allow its possibility. Indeed, it might even be that she always get every cup backwards; i.e. she is always wrong. This is as bland a set of premises as possible. In advance of the experiment, we will assume merely that she can get any number of cups right, from 0 to N.

We have our model in hand. "Has the ability" in our model says that the lady can guess any number of the N cups correctly. All the lady knows is that N is divisible by 2, that she will see an equal number of milk-first and tea-first cups. She will receive no feedback on her guesses (this is important). Thus, we do *not* assume (initially) she will employ an optimal guessing strategy.

What is an optimal guessing strategy? Suppose we gave the lady feedback and told her whether her guesses were right or wrong as the experiment progressed. If, say, the first four cups were all milk-first and she knew she got these all correct, even if she has no ability and did so just by guessing, then (if she was paying attention) she ought to get the last four correct, too (even before tasting!). My experience with ESP testing suggests most people do not use optimal guessing strategies, but if they did we can account for it, though it's not easy to do so. So for ease, we'll forbid feedback.

Question 1 Given this model (and *only* our other premises), and *before* running the experiment, what is the probability in our uncertainty the lady guesses 0 right, 1 right, 2 right, up to N right? This question is equivalent to asking what fraction of cups she will guess correctly: $0/N$, $1/N$, up to N/N. It is *not* equivalent to asking what sequence of correct and incorrect guesses she will evince. The fraction of correct guesses is easily answered, for $0, 1, \ldots, N$ is $1/(N + 1), 1/(N + 1), \ldots$; that is, the probability that she guesses j cups correctly is $1/(N + 1)$ for $j = 0, 1, \ldots, N$.

Stated yet one more way, since we have assumed as a premise the model that she may guess any number of cups correctly, the probability that she does so is 1 divided by the number of possibilities. (That last statement is derived in the previous chapter.)

Question 2 Suppose we run our experiment for $2 < N$ cups and are interrupted. Given our model and premises, but also given her guesses up to this point, what is the probability that she guesses 0 cups right, 1 right, up to $N - 2$ cups right? The *exact* answer has a simple mathematical form (as in the last chapter in developing parameters).

Question 3 The experiment is finished. She has guessed M correct out of N (M is a sum of the correct milk-first and correct tea-first cups). Given our model and given M, what is the probability that she guessed a fraction K/N correct, where K does not equal M? It is 0. A silly question to ask, yes, but let's expand it. Same premises: what is the probability she guessed a fraction M/N correct? It is 1. Another silly question, trivially answered. So why bother?

Ordinary hypothesis testing theory would have us ask something like this: what is the probability that she guessed $(M + 1)/N$ correct, and the probability she guessed $(M + 2)/N$ correct, and $(M + 3)/N$ correct, up to N/N correct? For us, and conditional on what happened, the sum of these probabilities is 0, as we agreed. But not in hypothesis testing, where the meaning of the word "guessed" is changed. It no longer means "guessed" but "*Might* be guessed were we to embed the experiment in an infinite series of experiments, each 'identical' with the first but 'randomly' different; we also hypothesize that if we were to average the correct guesses of this infinite stream, the result would be *precisely $N/2$ correct guesses*."

In other words, frequentist theory demands we calculate a probability of what *could* have—but *did not*—happen in fictional "repeated trials" (where "repeated trials" is shorthand for "embedded in a sequence of infinite repetitions"). The theory must also hypothesize a baseline, a belief that the infinite sequence converges to some precise average (here, $N/2$ correct guesses). Stated differently, frequentist theory asks the probability of seeing results "better" or "worse" than what we actually saw, given the model is true, a value for the baseline, and M.

This violates our agreement that we should use *only* the evidence from the experiment (and knowledge of the experimental set up) to test the truth of our model. Hypothesis testing does not make statements about what happened, but what might have happened but did not in experiments that will never be conducted.

This probability is the *p*-value, as is obvious. If the *p*-value is "small", the hypothesis that the baseline is $N/2$ is "rejected", i.e., it is believed to be, or rather *decided* to be, *certainly* false. The *p*-value does not give a probability that the baseline is false: it instead asks us to believe absolutely in the truth or falsity some contingent hypothesis (i.e. that "baseline $= N/2$"). In other words, a decision based on the *p*-value implies that the probability of "baseline $= N/2$" is 1 or 0 *and no other number*.

Harold Jeffreys famously summarized the predicament of hypothesis testing, "What the use of P [values] implies, therefore, is that a hypothesis that may be true may be rejected because it has not predicted observable results that have not occurred."

It is finally time to reveal what happened! Our good lady guessed $M = N = 8$ cups: she got them all right. (Some reports claim she got $M = 6$ right, missing one milk-first and one tea-first guesses; for our sake, it doesn't matter.) Remember our goal: we want to know whether or not she "has the ability." The hypothesis calculates this:

$$\Pr(T(M,N) \geq |t(M,N)| | \text{Does not have ability}), \qquad (9.8)$$

where $T(N)$ and $t(N)$ are some identical mathematical function of the data, but where the $t(M,N)$ is the value of the statistic we actually observed and $T(M,N)$ is the value of the statistic in repetitions of the trial, where these repetitions are embedded in an infinite sequence of trials.

$T(M,N)$ and $t(M,N)$ are called "statistics"; they are not unique; their form and use are not deduced. They are entirely *ad hoc*. Indeed, for this experiment we have (at least) our choice of the binomial and Fisher's exact statistics. For the former, (9.8) = 0.0039 and for the latter (9.8) = 0.014. We could have easily expanded this list to other popular test statistics, each providing different solutions to (9.8). Fishing around for a test statistic which gives pleasing results is a popular pastime (we want the statistic or statistics which give 0.05 or less for (9.8), this being the magic number).

Which of these is *the* correct test statistic? Neither. Fisher's test could be used if the lady knew she was getting exactly four cups of each mixture, the binomial could be used if she didn't; but other choices exist. (It is the lady's perspective that matters there, not yours.) In any case, we have two *p*-values. Can they help answer our original question? They cannot. Equation (9.8) is not Eq. (9.7). In no way is (9.8) a proxy for (9.7); it is even *forbidden* in frequentist theory to suppose that it is. Classical theory merely says that if (9.8) is less than the magic number we "reject" the theory "she does not have the ability". That is, we claim that "she does not have the ability" is false, which necessarily makes "she has the ability" true.

But recall that "she has the ability" had *multiple* interpretations. Which of these is the hypothesis tester saying is the right one? Well, none of them and all of them. Actually, the answer the anxious hypothesis tester will give when posed this question is usually a variant of, "Is that the bus? I must run." However, there is still the "agnostic" model; see below. Incidentally, if she got two wrong, (9.8) is 0.24 for Fisher's and 0.14 for the binomial.

We cannot answer (9.7) without first deciding what "She has the ability" means. If we decide, in advance, it means "She always guesses correctly" then as long as $M = N$ this theory has probability 1, i.e. (9.7) = 1. If $M < N$ then (9.7) = 0. And that is that. If we decide it means "She always guesses at least $N/2$ correctly" then as long as $M >= N/2$, (9.7) = 1, else it is 0. And similarly for any other interpretation.

That means that if we have one fixed interpretation and are willing to entertain no other, then as long as the observations are consistent with this theory, we must continue to believe this theory is *certainly* true (conditional on our premise). And if the evidence is not consistent, we will have falsified our interpretation and thus it must believe it is *certainly* false. But if we have falsified it, this does *not* mean we have given a boost to some other theory because, of course, we have already said that there were no other theories. This is a serious and fundamental point.

In order to have non-extreme probabilities attached to a model's truth, we must have more than one model in contention. One model alone is either true or false: the premise of only one model leads to the tautology "this model is correct", which is why this premise does not provide additional evidence. So suppose we have decided that "has the ability" means either M_1 = "always guesses correctly" or M_2 = "guesses at least $N/2$ correctly". Good arguments, after all, can be made for both interpretations. Before we see the experiment, based on these arguments, we must assign a probability either is true. If our evidence is *only* that we have these two to pick from, then we would assign probability 1/2 to each. This is weak and arbitrary. Far better to look outside our data and experiment to understand which of these is more likely, meaning finding premises which are probative of both. But *for ease only*, each equi-probable.

Then if we see $M = N - 1$ (which is $> N/2$) then we have still falsified M_1; this necessarily makes the probability of M_2 as 1 (given our premises). And if we see $M < N/2$ we have falsified both— *leaving no alternative*. But if $M = N$ then since this evidence is consonant with both models, we have not changed the evidence that either is true. This is it; this is *the* answer no matter how many interpretations we initially consider.

The one possibility left is the agnostic model. Suppose the lady got $M = 0$ right in $N = 40$ cups (say). Would you say she "has the ability"? Sort of: she appears to be a perfect negative barometer. If you knew somebody who was always wrong about picking stocks, he would be as useful to you as somebody who was always right. So we leave ourselves agnostic about her ability and say it could be guessing anything from 0 to N. At the end, we remain agnostic but we are able to predict how well she will do in N *new* trials. This is important because even if we are agnostic, there are different forms of agnosticism. That is, we are assuming uniform agnosticism, but it may be that a better model might be one which allows different performance for milk-first and tea-first cups. And it could be that milk-first and tea-first cups differ, but her palate fatigues after W cups. And so on and on for all the other possible models.

Being agnostic has *not* excused us from formulating a model—which we can test and verify on new data. We have turned hypothesis testing into science. But enough is enough. Never use hypothesis tests. That sounds dogmatic. Aren't there instances where we have an unobservable hypothesis which we would like to know is true or false and where "hypothesis testing" may seem reasonable? Take situations akin to jury trials, where the hypothesis of interest is "guilt", and where the "null" might be "not guilty". Evidence is accumulated and a decision is made, reject the null or fail to accept it.

But this is only a surface similarity: jury decisions are not like classic hypothesis test after all. What happens is this. The proposition of interest is "He did the deed" or "He is guilty". We want the probability of this with respect to the evidence presented in court and how that evidence relates to the common-sense wisdom of the jury. One piece of evidence might be a positive confession from the accused. "I did it," he said. We want Pr(Guilty|Confession & C) where C is the collective wisdom of the jury as it relates to this trial. C is that knowledge the jury brings with them coupled with the evidence, some of it intangible, discovered in court. For instance, in C might be prior suppositions about false or coerced confessions coupled with ideas about why the accused could not look anybody in the eye when confessing. Whatever Pr(Guilty|Confession & C) is, it won't be amendable to calculation. Evidence piles up. We want Pr(Guilty|Evidence at this time & C). Those who have ever sat on a jury knows that this probability swings in wide arcs, and not just because evidence accumulates, but because some is forgotten and then brought back to mind.

In any case, at the end of the trial we have $p = \text{Pr}(\text{Guilty}|\text{All Evidence \& C})$. This also is not quantifiable, and thank the Lord for that. The last thing we need is the scientism inherent in meaningless quantification made part of our legal system. But even though p isn't a number, jurors have an idea about it. That all don't agree is not because probability is subjective, but because the "All Evidence & C" is different for every juror (and even for the same juror through time).

In criminal trials in the United States at this writing (but things change), jurors are asked whether or not their p represents "guilt beyond a reasonable doubt." This does not imply any numerical value for p; again, thank God. But all are told it means a very "high" p. Now most juries have to be unanimous, which means each has to have a high p. These p will differ juror to juror, but only because they are working with different sets of evidence (which includes the common-sense). Discussion between jurors at this point begins, which is a process to agree on just what evidence is relevant, and this includes that introduced in the courtroom and that brought in by common-sense. If this works, then all are brought crudely to the same point, which, since probability is argument, means everybody's probabilities must be in rough agreement. It would be in perfect agreement if everybody agreed exactly on the evidence. Thus it is not the probability itself which is important, but the process whereby jurors agree on relevance and irrelevance. Juries decide fact.

Any probabilities which follow from this fact are not nearly as interesting, except that the p must be "large", after which a *decision* is made whether the collective-p is "large enough" to decide guilt. This decision, like all decisions, is independent of the probability. Different crimes and circumstances are accompanied by different decisionable p. A crime may be minor and the expected punishment so trivial that the jurors decide on a low (yet still "high") p to conclude guilt. Or again the crime moderate but the expected punishment so frightening that a much higher p is demanded. Whatever the decision, it is unrelated to the probability on the accepted evidence. Though some jurors might be tempted to exclude or include certain evidence to lower or raise the group p. Of course, if agreement cannot be had, the jury can be declared "hung", which might bring a warning to come to agreement on the evidence or a dismissal of the jury.

It is worth noting that some jurisdictions are adopting a "preponderance of evidence" criterion for some crimes, such as rape. This is an awful criterion, for consider if the *only* evidence we have is she said he did it, and he denies, the probability of guilt is exactly at 50%—a rare case where we can calculate a number (via the statistical syllogism). But since we are in this perilous state, it only requires the slightest additional evidence to push the probability greater than 50%, a "preponderance" of evidence, and so convict the man. This is awful because it is precisely in those jurisdictions which adopt this criterion that tend to eschew juries, relying instead on the decision of (an interested) "judge." It is far too easy for a judge in rape case to come to a probability larger than 50%.

Anyway, even here we do not have a hypothesis test in the classical sense. We have a probability, quantifiable or not, based on evidence. In other words, ordinary probability as argument.

Lastly, to wrap up the PM2.5 example, a standard test of proportions (z-test) gives a p-value less than the magic number, ordering the conclusion that PM2.5 is therefore a cause of cancer of the albondigas. We have already seen this ascription of causality is unwarranted. Amusingly, if one fewer person (14 instead of 15 of the 1000 in the high group) had cancer, the p-value is no longer wee and we are then forced to conclude PM2.5 is not a cause. So much hinges on one observation! But some will argue that a different test would give different p-values. This is true, and demonstration of yet another weakness of hypothesis testing. p-Values are taken up next.

9.7   Die, p-Value, Die Die Die

So much has been written on the dismal subject of p-values, including some above, that it seems like piling on to say more. Here are a few out of hundreds of critical articles, [19, 48, 49, 94, 119]. But I do say more only to note three things, two of which are commonplace and that lead to a third which is not.

First, nobody ever remembers the definition of a p-value. Everybody translates it to the probability, or its complement, of the hypothesis at hand. For this reason alone p-values should be abandoned. Second, even some self-labeling Bayesians want to keep p-values, but in a Bayesian sense. This is to give an old error a new name, but it will still be an error. Thirdly is something more interesting: the arguments commonly used to justify p-values are fallacies. Here is the proof.

It turns out that frequentist theory implies that the distribution of the measure of difference, like in the race-income problem, actually called the "p-value of the test statistic", is "uniformly distributed". What that means is discussed in the next section, but what the theory implies is (something like), "If the null is true, the p-value can be any number between 0 and 1, and is equally likely to be any of them". The argument people employ, however, progresses like this: "The null entails that we see a p-value between 0 and 1. We see a p-value that is less than the magic number. Therefore, the null is false, or rather rejected as if it were false."

This argument is not valid because the first premise says we can see any p-value whatsoever, and since we do (see any value), it is actually evidence *for* the null and not against it. There is *no p-value* we could see that would be the logical negation of "$0 < p$-value < 1"; other than 1 or 0, which may of course happen in practice.[1] And when it does happen in practice, then regardless whether the *p*-value is 0 *or* 1, *either* of those values legitimately falsify the null, not just 0. That is, an observed *p*-value of 1 is evidence against the null, according to the argument.

Importantly, the first premise to that argument is *not* that "If the null is true, then we expect a 'large' p-value," because we clearly do not. But the argument would be valid, and the null truly falsified, if the first premiss *were* "If the null were true we would see a large p-value," but nowhere in the theory of statistics is this kind of statement asserted. Though something like it often is. R.A. Fisher, the inventor of *p*-values, was fond of saying this—and something like this is quoted in nearly every introductory textbook, [76]:

> Belief in null hypothesis as an accurate representation of the population sampled is confronted by a logical disjunction: *Either* the null is false, *or* the p-value has attained by chance an exceptionally low value.

This is the same argument as before; but Fisher's "logical disjunction" is evidently not one, as the first part of the sentence makes a statement about the unobservable null hypothesis, and the second part makes a statement about the observable *p*-value. But it is clear that there are implied missing pieces, and his quote can be fixed easily like this: "*Either* the null is false and we see a small p-value, *or* the null is true and we see a small p-value." Or just: "*Either* the null is true or it is false and we see a small p-value."

Since "*Either* the null is true or it is false" is a tautology, and is therefore necessarily true and thus can be removed, we are left with, "We see a small p-value." Which is of no help at all. The *p*-value casts *no direct light* on the truth or falsity of the null. This result should not be surprising, because remember that Fisher argued that the *p*-value could not deduce whether the null was true; but if it cannot deduce whether null is true, it cannot, logically, deduce whether it is false; that is, it *cannot falsify* the null.

Current practice is that a small *p*-value is taken to be by *everybody* to mean "This is evidence the null is false or likely false." That is because people are arguing like this: "For most small p-values I have seen in the past, the null has been false; I now see a new small p-value" as evidence for the proposition "The null hypothesis in this new problem is false." But this doesn't work because the major premise is false, or at least unknown.

Given all this, and of the myriad other criticisms no doubt well known to the reader, plus the ineradicable Cult of Point-Oh-Five, it is far past the time for *p*-values to go.

Lastly, because confidence intervals are sometimes seen as the fix or alternative to *p*-values, let me prove to you nobody ever gets these curious creations correct. According to frequentist theory, the definition of a confidence interval (for a

---

[1]The simplest example is a test for differences in proportion from two groups, where $n_1 = n_2 = 1$ and where $x_1 = 1, x_2 = 0$, or $x_1 = 0, x_2 = 1$. Small "samples" frequently bust frequentist methods.

parameter) is this. If an experiment is repeated an *infinite* number of times, each one "identical" to the last except for "random" differences (ignore that this is meaningless), and for each experiment a confidence interval is calculated, then (say) 95 % of these intervals will overlap or "cover" the "true" value of the parameter. Since nobody ever does an infinite number of experiments, and all we have in front of us is the data from this experiment, what can we say about the lone confidence interval we have? Only this: that this interval covers the "true" value of the parameter or it doesn't. And that is a tautology, meaning it is always true no matter what, and, as we learned earlier, tautologies add no information to any problem.

We cannot say—it is forbidden in frequentist theory—that this lone interval covers with such-and-such a probability. And even if we manage to repeat the experiment some finite number of times, and collect confidence intervals from each, we cannot use them to infer a probability. Only an infinite collection, or rather one in the limit, will do. If we ever stop short and use the finite collection to say something about the parameter, we reason in a logical and not frequentist fashion. And if we use the length of an interval to infer something about the parameter, we also reason in a logical and not frequentist fashion. Since the majority of confidence intervals in use imply a "flat" (improper, usually) prior on the parameter of interest, all working frequentists are actually closet Bayesians. Now all we have to do is take the short step from Bayes to logic, and probability will be on firm ground everywhere.

9.8 Implementing Statistical Models

We know what not to do. So what do we do instead? The answer is in (9.1), but that cartoon equation is not especially helpful. It's right in spirit but not in detail. Here are more details, saving specific implementations until the final chapter. A paper on the final sections of this chapter is available at [29].

Let past observables be labeled $D = (Y, X)_{\text{old}}$, where Y is the observable in which we want to quantify or explain our uncertainty, and X are the premises or observables assumed probative of Y (the dimensions of each will be obvious in context). Let the premises which lead to a probability model (if one is present) be labeled M. And let $X = X_{\text{new}}$ be the premises or assumed values of new observables. The goal of all probability modeling is this:

$$\Pr(Y \in y | X, D, M), \tag{9.9}$$

where y are values of the observable Y which are of interest to some decision maker, and there Y is written to express the quantitative portion of the proposition Y (if it has one beyond true and false). Models should be rare, because most probability is not quantifiable—and we must resist the temptation to force quantification by making up scientific-sounding numbers. But even if we do, (9.9) can be calculated, as long we as supply the premises which led to our creations. Although it may be obvious, the equation reads, "The probability Y takes the values y given the premises or assumptions X, the past data D, and the model M." If the model is parameterized and Bayesian philosophy is adopted, (9.9) is the posterior predictive distribution, and M incorporates those premises or assumptions from which the priors are deduced (see e.g. [21, 184]). The key is that no parameters are explicit in (9.9); the uncertainty in them has been "integrated out." Only observables and plain assumptions remain. Logical probability would supply premises from which the model M is deduced (there would be no parameters thus no priors).

Equation (9.9) eliminates, or rather combines, the efforts of testing and estimation into one form. The focus is entirely on observables and the assumptions made and their effect on the uncertainty of not-yet-seen or unknown values of Y. Not-yet-seen values of Y are those unknown or assumed unknown; usually they are as yet unmeasured, e.g. in the future.

A simple example of a deduced model is a die roll in which M = "This is a six-sided object with labels one through six and which when tossed must show only one side." The model is deduced based on these premises. There is no X probative beyond M and D can be absent or can be a record of previous flips (i.e. X and D are null or are assumed not probative). An application of (9.9) is $\Pr(Y = 6|X,D,M) = 1/6$. Because the model was deduced, no parameters were ever present.

It is unfortunately rare that models are deduced; most are posited *ad hoc*. That is, M is usually "I'm using regression", i.e. an act of will. Model deduction can be accomplished if the measurement of observables are properly accounted for, as we saw earlier. Another deduced model of finite "successes" and "failures" was given in Chap. 8 in the discussion of parameters. That model gives the predictive probability of seeing so-many future successes and failures given we have seen this many thus far. Most of the time, however, we are stuck with arbitrary or capricious models. Suppose we are interested in Y = "First-year college grade point average" of students. Observations X thought probative are the high school grade point average and SAT score. The model M will be ordinary regression. The goal is to produce statements like this:

$$\Pr(Y > 3.8|X_h = 3.5, X_s = 1160, D, M), \tag{9.10}$$

where the subscript "h" is for high school GPA, and "s" is for SAT. The D are past observations. Since regression uses continuous normal probability, we unfortunately cannot ask about observables like "$Y = 4.0$" and must restrict our attention to intervals.

At any rate, the main questions of interest are only two: (1) how do changing values of X_h and X_s change the uncertainty in Y in the presence of D, M, and y, and (2) of what value or descriptive power is M? M includes X_h and X_s as components, in the sense that the premises that led to M led to deciding these objects were probative of Y. Since the beginning of this book it has been emphasized that the propositions and premises we use are decided by us; it is still true in statistical modeling.

There is no notion of "significance" in either of these questions, and no notion of correctness in any instance where the model was not deduced. All probability is conditional, and that means the probabilities given in equations like 9.10 are correct. If, at some later point, it is decided that X_S is of no interest and the model is updated to reflect this, then the probabilities derived from the new model are also correct. One model may be superior to another, however, but only with respect to decisions made conditional on the models.

There is more work to be done by model builders in this way because values of y must also be chosen, and so do values of X. This implies a model may be useful in some decision contexts and of no use or even harmful in others. There is and should be no default or automatic levels of usefulness. The last thing the field of probability and statistics needs is another magic number *à la* "significance" with p-values.

The importance of probative observables and assumptions X is thus also a matter of decision and cannot be made automatically. This is plain from Eq. (9.9), which encapsulates *all* we know about Y given our assumptions. It is we who made these assumptions, and we who can change them. Again, the only ultimately true model of Y is that which is deduced from true premises, as in the die and success-failure examples (from the previous chapter). Every other model is therefore only useful or not conditional on the premises we assume. Since most models are *ad hoc*, as regression always is, we can only speak of usefulness. Deduced models are true by definition and thus nothing more need be done with them except make predictions. Deduced models do not even need to be verified.

The idea is, conditional on D and M, to vary X in the range of expected, decisionable, or important values to some decision maker and see how these change the probability of $Y \in y$. If a particular X as it ranges along the values we choose do not change the probabilities of $Y \in y$ in any important way, then these X are themselves not important. The opposite is also true. Importance is a matter of decision, which varies by decision maker. Importance is not a probability or statistical concept and therefore cannot be ascertained within probability models. If the probability of $Y \in y$ changes in any as X does, then that X is relevant to understanding Y, else it is not. Relevance is a probabilistic concept, but as the reader will see, it is almost always present given the assumption that the X is causally related to the Y. If X is known not be causally related to Y, then X is irrelevant by definition, and therefore should *not* be part of any model.

There is no hypothesis testing in the frequentist or Bayesian sense (as implied by Bayes factors, for instance). And there is no estimation of parameters. There are only plain, understandable, and verifiable probability statements. These probability statements can and should and must be verified. This allows communication of model goodness and usefulness in an intelligible, actionable manner. It reduces over-certainty but cannot eliminate it unless models are deduced.

Here is an example. There are 100 observations of first-year college students' grade point averages. We want to quantify the uncertainty in the GPAs of new students given these observations, and also given information thought probative,

in this case high school GPAs and SAT scores.[2] We assume an *ad hoc* ordinary
regression model. If we adopt the Bayesian philosophy, we need priors, and here an
assumption of "flat" priors will do. As is well known, in ordinary regression this
assumption matches the answers given by frequentist philosophy. But it doesn't
matter. Any premises that give different priors will do. Our purpose here is
not (directly) to investigate priors, but the uncertainty inherent in Y given the
assumptions we make.

It turns out

$$\Pr(Y > 3.8 | X_h = 3.5, X_s = 1160, D, M_{h,s}) = 0.038. \tag{9.11}$$

Notice that the dependence of the model on the assumptions has been annotated,
as it *always should be*. If, given D, we insist on M and on the presence of X_h and
X_s, then this is the final and true answer. Nothing more need be done. The values
picked for y, X_h, and X_s are those I, and perhaps nobody else, thought important.
A different decision maker might pick different values.

But suppose I am interested in the relevance of X_h. Its presence is an assumption,
a premise, one that I thought important to make. There are several things that can be
done. The first is to remove it. That leads to

$$\Pr(Y > 3.8 | X_s = 1160, D, M_s) = 0.0075. \tag{9.12}$$

Notice first that *both* (9.11) and (9.12) are *correct*, they are both conditional truths.
The probability in (9.11) is five times larger than in (9.12). This is a measure of
relevance and importance, given y = 3.8 and $X_s = 1160$. Importance and relevance,
like probability itself, are always conditional on our assumptions. A second measure
of importance is the change in probabilities when X_h is varied. That can be seen in
the following figure (Fig. 9.1).

There is a change from about 0 to 8 % over the range of high school GPAs. If
high school GPA was not probative of $Y > 3.8$ given these premises then the graph
would be flat, indicating no change. In other words, it would resemble the dashed
line, which is (9.12), the model without high school GPA. Is this "departure" from
flatness important? There is no single answer to this question. That entirely depends
on the uses to which this model is put. If a decision would be made differently
given these varying values of the probability, then high school GPA is important,
otherwise it is not. The answer is not a matter of probability or statistics. That the
line is not flat is *proof*, however, that, given M, y, and X_s, knowledge of high school
GPA is *relevant* to knowledge of Y.

It cannot be emphasized too strongly that importance and relevance are conditional, just as probability is. A linear function of high school GPA added to a
regression model already supplied with high school GPA would be irrelevant. It
might be that high school GPA is relevant or important at some levels of SAT, and

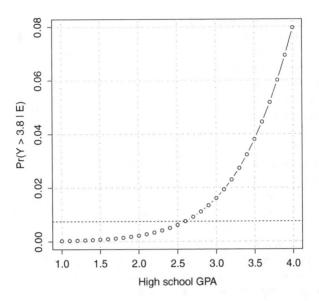

Fig. 9.1 $\Pr(Y > 3.8|X_h = x, X_s = 1160, D, M_{h,s})$ allowing X_h to vary from 1 to 4 in increments of 0.1. The notation on the figure conditions on E, which is shorthand for all the evidence we have. The dashed line is $\Pr(Y > 3.8|X_s = 1160, D, M_s)$

Fig. 9.2 $\Pr(Y > 3.8$ $|X_h = 3.5, X_s = 1160, X_w = x, D, M_{h,s,w})$ allowing X_w to vary from 0 to 26. The notation on the figure conditions on E, which is shorthand for all the evidence we have. The dashed line is $\Pr(Y > 3.8|X_h = 3.5, X_s = 1160, D, M_{h,s})$

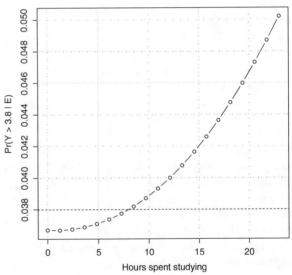

irrelevant and unimportant at others, and the same is true for the goodness measures of SAT. Information X is not isolated and is related to all the assumptions we made, and these include the other X in the model.

Now add the information X_w = "hours studied a week" to the model and create a relevance plot for it (Fig. 9.2).

The (conditional, as always) probability of $Y > 3.8$ varies little, from 0.038 to about 0.044, a change of only 0.006. Time studying is relevant because the probability differs from the straight line, which is the probability using the model *sans* time studying. But it is important? Given the values of high school GPA, SAT, and y, the change from 0.038 to 0.044 is small and therefore many decision makers might conclude that adding time studying provides no benefit to our understanding. Of course, such a small change might be useful to somebody. It is not the statistician's job to decide, but the decision maker's.

I cheated, because "time studying" isn't that at all. In fact, it is made-up numbers using R's rnorm() and abs() functions. It is therefore not surprising that the pertinent probability should vary little. That it does at all is only because of coincidence. By "coincidence" I do not mean "randomness" or "chance", because I know the determinative cause of these numbers, but happenstance, a known lack of causal connection between the made-up numbers and the observable of interest. We should add information presumed probative into a model *only* if we have a plausible belief that the information is related (somehow) to the cause of the observable of interest. If this connection is lacking, the information should not be added. Thus "time studying" should certainly be removed.

The strategy is to create scenarios that are of direct interest to a decision maker, the person or persons who will use the model. Plots like those above can be made at the values of the probative observables in which the decision maker is interested. There is no one set of right or proper values, except in the trivial sense of excluding values that are, given exterior information, known to be impossible. For instance, given our knowledge of grade points, the value $X_h = -17$ is impossible. Assessing relevance and importance for large models will not be easy. But who insisted it should be? That classical statistical procedures now make analysis so simple is part of the problem we're trying to correct.

Given the model and old observations, every set of X_h and X_s, at some y, produce a prediction. For instance, for future students with $X_h = 3.5$ and $X_s = 1200$ the (conditional) probability that $Y > 3.8$ is 0.045. In regression, incidentally, we do not have to restrict ourselves to a fixed y, because the model will produce a prediction of every possible value of Y. These predictions can and *must* be verified. An example of such a report is given in Table 9.1. Considerable art and thinking will have to go into presenting predictions from a model loaded with probative X. This may seem like a drawback, but in fact it is a boon. Far too many models are crammed with extraneous "controlling variables"; their usefulness is scarcely ever considered. This approach forces such consideration and encourages leanness; which is to say, models without fat. Lean models are not necessarily small. This example will be concluded after model goodness is introduced in the next section.

Other authors are beginning to advocate similar approaches, such as [152], who recommend "blind" testing of results. The predictive approach does away with the necessity of blinding, because the results are there for all to see. Of course, no method of release of results can forestall cheating or authors fooling themselves, but the predictive approach minimizes the harms from such activities.

Table 9.1 The predictions for $\Pr(Y < 2|X_h = h, X_s = s, D, M)$ and $\Pr(Y > 3|X_h = h, X_s = s, D, M)$ for common values of h and s, and two points of y thought to be of interest

	$\Pr(Y < 2|h, s, D, M)$				$\Pr(Y > 3|h, s, D, M)$			
h/s	400	800	1200	1600	400	800	1200	1600
0.5	0.99	0.94	0.760	0.4600	3.9e-05	0.00065	0.0093	0.071
1.0	0.98	0.89	0.640	0.3400	1.4e-04	0.00180	0.0200	0.120
1.5	0.95	0.81	0.510	0.2200	5.0e-04	0.00500	0.0420	0.190
2.0	0.90	0.70	0.370	0.1300	1.7e-03	0.01300	0.0820	0.290
2.5	0.83	0.57	0.250	0.0730	5.0e-03	0.03100	0.1500	0.420
3.0	0.73	0.43	0.160	0.0360	1.4e-02	0.06600	0.2400	0.550
3.5	0.61	0.30	0.089	0.0170	3.3e-02	0.13000	0.3700	0.680
4.0	0.48	0.20	0.048	0.0074	6.9e-02	0.22000	0.5000	0.790

Every X in the model appears in the table

9.9   Model Goodness

All probability models have a predictive sense. Equations (9.1) and (9.9) are written predictively, but even classical parametric models imply (9.1) or (9.9) once the parameters are eliminated ("integrated out"). Statistical models say Y will happen, or will be revealed to us, with a non-extreme probability. Physical models often claim certainty of Y, though almost always users of these models add "fuzz" to the predictions to avoid falsification, as we have seen. A weather forecast model might say Y = "Today's high will be 72°F" and if we observe 71 °F the model is formally falsified, but most will say "close enough". Whether they say this will depend, of course, on the decision made conditional on the model. The weatherman who issued the 72 °F would have better served his clientele had he attached uncertainty to the forecast. All forecasts of the contingent should ideally be stated probabilistically. Removing probability, neglecting the state the uncertainty, *makes a decision* and that decision may not be, and probably won't be, best for all users of the forecast. Eliminating uncertainty makes a decision because it transforms the probability into a bet. And not all would make the same bet.

Any model which issues non-extreme probabilities can be never proved wrong; that is, it cannot be falsified *unless* it gives precisely zero probability to an observation that is later seen, e.g. [3, 6, 8, 159, 160]. Given most probability models use or assume continuity, falsification never happens. But models can be shown poor in comparison with rival models of Y. A model can also be classed as good or bad based on decisions made with the model. Indeed, every (non-deduced) model should be judged by the decisions made with it. The same model can be valuable to one person and a burden to another because these people make different decisions based on the model. Because of this fluidity, model verification is a broad subject and can't be covered in detail here. But some general comments can be given.

First, what to verify? There are only two possibilities: the data already seen and that which is yet to come (or be made known to us). The traditional terms are in-sample and out-of-sample data. What usually happens in building statistical (or even physical) models is a model-fitting process. Premises are put forward which imply a model, and that model is judged or weakly verified with a generally ill-thought-out or harmful measures such as *p*-values, and then premises are added or rejected in an effort to make the model "fit" the observed data better. The process stops at some point. The measures of fit of this final model *to the sample at hand*, like *p*-values or "residuals", are put forward as "proof" of the model's veracity. That's when (in some fields) explanations start to fly thick and fast. After all, some thing or things caused the data at hand, and if the model "fits" well the hypothesis-testing false-dichotomy (written about above) is invoked and theories aplenty are discovered to be *the* cause. Sometimes caution is urged and it is reluctantly admitted that the theories bruited might be mistaken but this is mere boilerplate, like adding "more research is needed" to every paper ever published, and in any case, theories promulgated are believed.

Under-determination, as we have seen, guarantees perfect (or necessary) certainty is not to be had for the contingent. Nevertheless, models can be useful epistemological devices. And they do provide evidence for causes, or are reasons to believe certain causes. Fitting past data well is certainly to be applauded and a clue that the model *might* have identified important associations of the proposition of interest. But past fit is weak, very weak, evidence of model goodness. As the saying goes, past performance is no guarantee of future success. This logical truth is nearly everywhere forgotten in the rush to *do something*. The appalling demand to produce ever more research is directly responsible for a great deal of nonsense put forward as "proved" because of classical statistics. And this is done in good faith. Everybody believes in confirmation bias, just as everybody believes it only happens to the other guy.

The true and only test of model goodness is how well that model predicts data *never* before seen or used in *any* way. That means traditional tricks like cross validation, bootstrapping, hind- or back-casting and the like all "cheat" and re-use what is already known as if it were unknown, they repackage the old as new. The key and overwhelming advantage of the predictive approach is that it displays models in the form where they can be used to make predictions *independent* of the model authors. Others can see for themselves whether the model "works." What can be fairer than that?

This approach is of course standard practice in some fields. Engineers who only build theoretical (model) bridges do not find large audiences. Classification models like, say, retinal or hand-writing recognition are an example of field-tested (often hybrid probability-physics) models—even though they go under names like "machine learning". Physicists once swore by the predictive approach, but some are lamenting this is no longer the case, see [68]. What theory (before the latter part of the last century) was ever believed without finding experimental support? But many subfields have voluntarily or under political pressure decamped Reality for the calmer pastures of Pure Models. The effects this has had are easily seen (by those left behind).

Every field that does not insist on out-of-sample verification has, because of classical statistics, suffered. Over-certainty is pandemic. We mustn't forget that the largest users of statistical models are not academics, but civilians and bureaucrats playing with spreadsheets, folks eager to arrive at a pre-determined destination. To restore sanity to science, it must everywhere return to predictive methods. That means insisting on models which speak of observables, which are testable. The best test, as you are by now tired of hearing, is how well the model works for decisions you yourself make with it. How the model works for others is always guesswork, but some statements are possible.

Testing in a general sense—when the specific decisions to made with the model aren't known—can be done with scores. A score takes two inputs: a probability for some Y and the observation of that Y. It provides a measure of closeness of the probability and outcome. A score treats the model as a model and not a decision. That is, the model is a probability forecast and not a decision or a rule based on that forecast. Usually scores are linear over forecasts, but they needn't be; dynamic scores are more than possible. Verifying time series predictions are an example. Another: a store manager buying quarterly stock using forecasts as guidance. Scores also needn't be singular, but can be multidimensional. Verifying meteorological, climatological, and even astronomical models come to mind.

Scores are also limited where decisions are not. A score would say the stock broker who always says the stock will rise when it falls and vice versa is terrible, but the opposite is obviously true. The predictions of that broker can be useful. Yet models must always be taken as they are given and not transformed in any way. Any, even the slightest modification to any model which is not deducible from the model premises or is not tautological makes a *new* model. Given my experience, this point cannot be stressed enough. A "model" which undergoes almost constant revisions is often called the same model. That might be useful for bookkeeping, but it's not a good principle in verification.

Let $f = \Pr(Y|X)$, a number. If Y is multi-valued, $F_y = \sum_y \Pr(Y_y \leq y|X)$ is obviously conditional on X, and Y_y is the schematic proposition "The value is y". And let O_y be the (ultimately) observed value of Y; O is a proposition, e.g. O_y = "The value y was observed". Since it helps to have an example, a popular loss function is the continuous rank probability score (CRPS), which is more or less a distance between the empirical distribution function of the model and the staircase step-function of the observable [95, 112]. Numerically:

$$\text{CRPS} = \sum_y \left(F_y - \text{I}\{y \geq O\}\right)^2 \qquad (9.13)$$

where I is the indicator function. The "continuous" part of the name is because (9.13) can be converted to continuity in the obvious way. Briefly, the score is proper, and it is sensitive to distance (meaning that observations closer to models score better). *Proper scores* are defined conditional on X. Given X, the proper probability to announce is F_y, but other probabilities could be announced by scheming modelers, say G_y, where is tacit on hidden premises likely related to X. Propriety in a score is when

$$\sum_y S(G_y, O)F_y \geq \sum_y S(F_y, O)F_y. \tag{9.14}$$

In other words, given a proper score the modeler does best when announcing the probability implied by X. Propriety is a modest requirement, yet it is often violated. The bearer of bad news often substitutes G for F knowing S is improper. Forecasts of doom, especially environmental doom *à la* Paul "Population Bomb" Ehrlich, at odds with the best evidence are usually more welcome than predictions of *status quo*. Incidentally, it is often said proper scores are those that encourage modelers to give the "true" probability of Y. As we know by now, there is no true unconditional probability. There is only that probability implied by the X which is *stated* or *claimed* to be held. Whether X really is or not held is always a mystery. In any case, the stated model should be (seen to be) deduced from X.

A perfect prediction has a CRPS (or any proper score) of 0, an interesting tidbit because it shows that models which do not put extreme probabilities on events can never be perfect. And models of the contingent should never assign extreme probabilities. Once again, perfection is not ours to have. The CRPS has found widespread use in meteorology, for example [223]. Other measures are possible and, as it cannot be stressed strongly enough, the best measure is that in accord with the decisions made with the model.

If F (as an approximation) is a (cumulative) normal distribution, or can be approximated as such (be careful: an approximation of an approximation!), then the following formula may be used:

$$\mathrm{CRPS}(N(m, s^2), O) = s \left(\frac{1}{\sqrt{\pi}} - \frac{O - m}{s} \left(2\Phi \left(\frac{O - m}{s} \right) - 1 \right) \right)$$
$$- s \left(2\phi \left(\frac{O - m}{s} \right) \right) \tag{9.15}$$

where ϕ and Φ are the standard Normal probability density function and cumulative distribution function, and m and s are *known* numbers.

CRPS, or any score, is calculated per prediction. For a set of predictions, the sum or average score is usually computed, though because averaging removes information it is best to keep the set of scores and analyze those. A few techniques for that will be given. But first we need the idea of *skill*. Skill is when one model demonstrates superiority over another. Skill is thus conditional on the chosen score. *Skill scores* K have the form:

$$K(F, G, O) = \frac{S(G, O) - S(F, O)}{S(G, O)}, \tag{9.16}$$

where F is the prediction from what is thought to be the superior or more complex model and G the prediction from the inferior. Since the minimum best score is 0 and given the normalization, a perfect skill score has $K = 1$. Skill exists when $K > 0$, else it is absent. Incidentally, receiver operating characteristic (ROC) curves are not to be preferred to skill since these do not answer questions of usefulness in a natural way; see [31] for details.

Models which do not have skill should never be boasted of. Why? Except in the most trivial sense, natural comparison models always exist. Take ordinary regression as an example (assume for now all parameters are known). Predictions made from a complex model F, say one with a dozen "regressors" (I'm thinking of economics or sociological models here) can always be compared with the "null" model's predictions G, i.e. a regression with no regressors, one which says the uncertainty in Y always and everywhere is quantified as a normal distribution (and the same distribution everywhere). F ought to be able to beat G. If it can't, F stinks, to put it in purely philosophical terms. As before, skill can be demonstrated in- or out-of-sample. Out-of-sample is the *only* true test of model goodness. Skill, like proper scores themselves can also be averaged or otherwise summarized across observations or examined with respect to model premises to assess where models do better and where poorly.

If it isn't already plain, every statistical model should be put into predictive form and checked in-sample by proper scores and for skill, and then re-checked at sometime in the future when new data becomes available for out-of-sample skill. Just as how engineers and physicists of old built their models.

Model calibration is also of interest in determining where and how a models perform. A *calibrated* model lines up with reality in its particularities. Calibration only concerns collections of forecasts and not individuals ones, as will be clear. I'm discussing actual and not theoretical calibration as is written about in for instance Dawid [52–54], which concerns what happens to models in the limit (which is never reached in practice; incidentally, though Dawid's prequential principle shares many things in common with full-conditional probability and the predictive approach, prequential statistics errs on cause and implies, at times loosely and at others more strongly, probability is a cause, where probability reaches back from infinity to the present "generated" values). These are interesting mathematical discussions but are of little practical value. These discussions invoke "nature" "picking" probabilities or make claims that there are "true distributions" meaning unconditional distributions, or rather that probability is ontological. "Nature" causes things to happen; and even in physics where we have excellent grasp of premises which *imply* certain probability models, it is that calibration *follows* from the model, as a property of the model, and is not the other way around, that somehow models "strive" for calibration. This is essentially the frequentist fallacy that gets things backwards, that defines probabilities as that which happens in some limit, rather than the deduced probability which implies the behavior of frequencies.

Calibration has three aspects: calibration in probability, exceedance, and marginal calibration, [96]; see also [151]. The model, conditional on various X, makes predictions. If the X were constant the model's predictions would also be constant. Each prediction, conditional on each X, is a separate (but fixed, deduced, predictive) number for each value or level Y can take, i.e. $F_{x,y}$. Next collect n observed values O for all predictions $F_{x,y}$; let O_y be the observed value of y, and form $\sum_y I(O_y \leq y)/n$, the so-called empirical cumulative distribution of the observable. This is usually written without its condition on X.

Calibration in probability is when for each x, y in which $F_{x,y}$ is constant (the same probability), the average of $\sum_y O_y/n$ for each of the y at that probability is equal to $F_{x,y}$. In other words, for those times the model says the probability of at least y is some probability, we want the observed frequency of values of at least y to equal that probability. So-called calibration plots are made to display $F_{x,y}$ by $\sum_y O_y/n$, but these plots are named for calibration in probability and not the other kinds. A model may be calibrated for some x, y and miscalibrated for others. Mark Schervish has proved (for models of a general kind) that models calibrated in probability are better in terms of scores than other similar models, [191]. The nature of his discussion would bring us too far afield here. Suffice to say that calibration in probability, and in the other ways, is desirable, as intuition suggests.

Over-confidence is the typical symptom of miscalibration in probability. Doctors, for example, are infamous of being too sure of themselves and, as a hypothetical but realistic example, that when one collects the instances when the doctor says "The probability of dying within six months is 80%", one usually finds only 20% of these patients have actually died. Phil Tetlock has made a career showing how over-confident political experts are in this sense, e.g. [211]. Consistent miscalibration in probability is good to be recognized because it is possible, but rarely done, that the miscalibrated model can be improved with simple adjustments. Of course, the loss one faces when announcing a probability may not be symmetric. That is, the loss for a false negative may be substantially larger than for a false positive, as is often the case. There are ways of incorporating decisions and losses into calibration and other verification, but these are too far afield for us here. For more on verification see [30, 31, 159, 160, 202, 220]

To understand exceedance calibration, first take the collection of $F_{x,y}$ for a fixed y (different x will give different probabilities for this fixed y). Each of these probabilities will equal a frequency from $\sum_y O_y/n$, where each frequency is possibly a different y. If the mean of these observed y equals the fixed y for all y, we have exceedance calibration. These too can be plotted.

Marginal (sometimes called climatological) calibration is easiest to obtain. This is when the mean of $F_{x,y}$ for a fixed y equals the mean of $\sum_y O_y/n$ for the same fixed y, and that this equality holds for all x, y. If Y were dichotomous, then the average of the predicted probabilities (averaged over all x) would equal the mean number of occurrences Y was true.

Lastly, it is worth emphasizing that a scenario is a projection is a forecast. Since all probabilities are conditional, and a prediction from a statistical model is conditional on x, we might just as well say that x is a "scenario." This is highlighted because of a disquieting tendency that has developed in some quarters to dismiss criticisms of failed models because model owners say their predictions were not forecasts or predictions but scenarios.

Let's now finish the GPA example started above. The verification strategy is this. A model is built using importance and relevance, as above. It is then released into the wild, as it were, to wait for new data to arise. Every new data point will have a value of (X_h, X_s, Y). These are fed into the model and the prediction for the probability of Y is given (with, say, $y = Y$). These prediction-observable pairs are then evaluated in relation to a proper score and possibly also with respect to a simpler version of the model, in a move to assess model skill. Probability leakage is also discoverable [26]. This is when the model gives non-zero probability to values of Y which are known to be impossible given external evidence, evidence which is usually ignored in the model-building process (regression, for instance, is horribly over-used). For instance, we learn that at this college a GPA of 4 is the maximum. Yet in our model with $X_h = 4$ and $X_s = 1400$ the probability that $Y > 4$ is 0.105. This is substantial leakage and a guarantee of model weakness.

We can also compare our touted model with a simpler model, which is perhaps a standing competitor or one otherwise natural to consider. In the example, time spent studying was revealed to be faked. But suppose I were to give the data to a statistician and not tell him of the fraud. Time studying sounds plausibly causally related to grade point. The check of relevance and importance do not excite. Thus it is reasonable to say two models are in contention, $M_{h,s,w}$ and $M_{h,s}$. If, considering whatever proper score we are using, $M_{h,s,w}$ cannot "beat" $M_{h,s}$, $M_{h,s,w}$ is said not to have skill in relation to $M_{h,s}$. In regression, as said, the so-called null model (only with an "intercept") is always available as a comparator.

As above, the CRPS is used. Skill is a relative measure of improvement this score, here $(\text{score}_{\text{full}} - \text{score}_{\text{partial}})/\text{score}_{\text{partial}}$, comparing a full or larger model with a partial, less complex, or otherwise natural comparator. The mean CRPS for the "full" model with "time studying" is 0.0734, while for the "partial" model without it is 0.0749. The skill score is 0.02. This shows the full model is superior, but only just barely. Here is a per-observation analysis of CRPS and skill.

The first graph in Fig. 9.3 shows the CRPS calculated for each observation for the full and partial models; a one-to-one line is over-plotted. The addition of "time studying" does not lead to uniform improvement. The next graph are histograms of CRPS scores; the partial model is the dashed line. This is useful for a decision maker deciding how valuable either model is. Again, the best score is that which is related directly to the decisions made with the models. Whether CRPS is this score is situation dependent. The last graph shows the skill score for each observation of college GPA. A mixed bag, as far as "time studying" goes. If we didn't already know, we'd suspect that "time studying" is not adding much, and is even subtracting from, our knowledge of college GPA.

In Fig. 9.4 show the skill plots over each X in the model for each observation. Skill is had for values greater than 0. The view that "time studying" is nearly useless is confirmed. The last graph is particularly revealing. As "time studying" moves to either extreme, the skill bifurcates, showing a process that "can't make up its mind." If "time studying" were truly valuable, the signal would be coherent. At this point, conferring with the decision maker, the statistician might drop "time studying" and compare his new "full" model consisting only of high school GPA and SAT with (perhaps) a "partial" model consisting only of an "intercept".

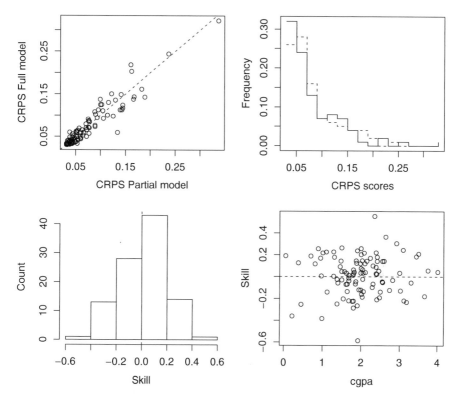

Fig. 9.3 *Top left*: the CRPS calculated for each observation for the full and partial models; a one-to-one line is over-plotted. *Top right*: the histograms of CRPS scores; the partial model is the dashed line. *Bottom left*: the skill score for each observation. *Bottom right*: the skill plotted for each observation of college grade point. Skill is had for values greater than 0

The verification process can be done on the already-observed data as an initial check on model goodness. The analogy is standard residual analysis. The same weaknesses apply, however, because the temptation to tweak the model to produce better verification measures will not be able to be resisted. Over-fitting and over-confidence will result, as always, but with an important twist. Since the model will be known and published in its predictive form, outsiders do not have to trust the in-sample verification. They can wait for new data and apply verification on them themselves.

Once verification measures are known, it is a mistake to say that X is linked to, or is associated with, or predicts Y, or, worse, some variant of "When *X* equals *x*, *Y* equals *y*". These are versions of a colossal misunderstanding, which is to say X *causes* Y, It is true that X determines the uncertainty we have in Y, but *determines* is analogical; it has an ontological and epistemological sense, as we now know. Probability is only concerned with the latter usage. The only function probability

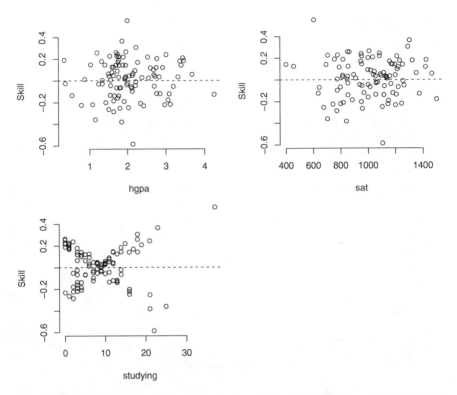

Fig. 9.4 *Top left*: the skill plots for each observation of high school GPA. *Top right*: skill for each observation of SAT. *Bottom left*: skill for each observation of "time studying"

has is to say how our assumptions X determine epistemologically the uncertainty we have in Y. If we knew X was a cause of Y, we have no need of probability.

Importance, relevance, skill and the like are replacements for testing and estimation, but not painless ones. The recipient of an analysis is asked to do much more work than is usual in statistics. However, this is the more honest approach. The benefit is that Eq. (9.9) answers questions which are always asked us in the form expected. The probabilities are in plain English and painless to interpret. Everything is stated in terms of observables. Everything is verifiable. The conditions on which the model relies are made explicit, made bare for all to see and to agree or disagree with. Gone is the idea that there is one "best" model which researchers have somehow discovered and which gives unambiguous results. Gone also is the belief that the statistical analysis has proved a causal relationship.

The model is made plain so that all can use it for themselves to verify predictions made with it. Everybody will be able to see for themselves just how useful the model really is.

9.10 Decisions

I have avoided the topic of decisions, except to insist on what is true: probabilities are not decisions, and vice versa. What somebody does with a probability is of no interest, so to speak, to the probability itself nor its calculation. This view is *contra* [18, 75, 190] and others who equate probabilities and decision. There are clear ties between decisions and probability, of course, as these authors also recognize; also see [23, 121]. The difference between probability and decisions was understood as early as Newman [161], who separated understanding of uncertainty into three progressive mental steps: doubt, inference, and assent. In our language, these are the forming of the proposition of interest and discovery of probative premises, objective probability calculation (numerical only if warranted), and then decision. Newman was writing prefatory to examining evidence of religious beliefs, so his book did not become well known in the probability literature.

To make a decision means understanding the uncertainty. Uncertainty is conditional on the assumptions made. But so are the gains and losses to be realized in any decision conditional on assumptions. Making a decision means also making a decision about how one wants to decide, a screwy but true statement. Does one want to avoid maximum loss? Does one want to realize some expected gain? This is an entirely separate field of study for which there is no room in this present volume, except to say one or two small things.

Many of the same admonitions made about probability apply to decision making. All decisions are conditional, just as all probability is. Change the premises, change the decision. All decisions involve finite and discrete objects, just like probability when it comes to real measurement of observables. Money, for instance, is not infinite nor continuous. Pseudo-quantifications, like "utility" often, ought to be avoided to keep the danger of the subjective fallacy at bay. A decision useful or optimal for, say, your doctor or government isn't necessarily optimal for you. And so on.

Clearly decisions are important to "model-building." What "variables" to include or exclude from a model are decisions, and based on the probabilities output from the model. Adding an X to the model changes the probability of interest by some small amount. Is that amount important? That depends entirely on what's important to the decision maker. No general guidelines can—*nor should*—be given. People like easy decisions, and the classical methods of statistics, using either p-values of Bayes factors, were designed to make decisions as easy as possible. People no longer had to think; hypothesis tests did the thinking for them. I'd like to see a return to difficult decisions, especially in models for complex entities, like human behavior, or large-scale biological or natural processes. To make a quick or easy decision is to risk over-certainty. And that means bad decisions.

It should also be emphasized that algorithms that lead to point-predictions, like, for instance, machine learning, "tree" models, and the like are making decisions and not (necessarily) providing probabilities. It's true they're basing these decisions on probability, but the results still eliminate uncertainty. This is fine if these models are "tuned" to decisions in particular contexts, usually in automated systems, like license plate number recognition, grocery store checkout scanners, and so forth (everything is put into a discrete "bucket" here, which is in line with the discrete limitations of measurement mentioned above). But if these models are meant for a broad or a scientific audience, they do that audience a disservice by not including uncertainty.

Frequentist modeling focuses on "likelihood"-based methods, which use parameterized models and where speaking of uncertainty in the parameters is forbidden, though guesses of the parameters are allowed (we've already seen confidence intervals are of no use in stating uncertainty). Information-theoretic methods which use plug-in parameter guesses are sometimes used to decide between models. The idea is that models which fit the data better while also penalizing model complexity are said to be "good" or better. The penalties for model complexity are needed because a model can always be found which fits any set of data perfectly. These "perfect" models will often be extraordinarily complex, and when they are used to make predictions of unknown measurements they fail spectacularly. The approaches of [34, 136] best illustrate these ideas. But they are not to be recommended. Firstly because the focus of modeling should never be on parameters, as demonstrated above and because parameter-centric models never fully state the uncertainty of a problem. But mostly because information theoretic model choice methods are like any other "bootstrapping" procedure which attempts to get something for nothing. Model fit is a necessary but far from sufficient criterion for model goodness. The *only* way to tell if a model is any good, as I have repeatedly stressed, is to use it. Of course, information theoretic techniques can certainly help in comparing full models and in giving an assist in characterizing parameter uncertainty (should one insist upon their use), as often demonstrated in Jaynes [122].

With those preliminaries, here are a few close intersections of probability and decision without having to discuss the nature of decision too much. The first is the gambler's fallacy. Black has come up often on last several spins of the roulette wheel, therefore, thinks the gambler, red is "due." He therefore bets red; i.e., he makes a decision based on his vague probability. This simplistic belief brings a smile to our face when we realize we're much smarter than the poor fellow donating his paycheck to the casino, but we shouldn't be so glib. What this situation different than the thousands of textbook examples illustrating suspicions of "biased coins?" (those same books will also have the gambler's fallacy). The answer is: nothing. The gambler thought probability was some kind of cause, just as probability is often thought to be the cause of coins landing this or that way.

Why do we think it is a fallacy to say "red is due"? Only one reason: because we understand the nature of the causes of the ball dropping, and we believe those causes—for there are obviously many on each spin—have remained constant in nature from spin to spin. Indeed, it is because we understand the nature of the causes, at least in a broad sense, that we can *deduce* the probability of red (and black and green). Notice that we do not need to know every cause in every detail, which is (as I demonstrated earlier) an impossibility if we include the primary cause. All we need to do probability is to have some knowledge, but it needn't be complete.

Very well, what if we suspect the wheel is crooked? Hypothesis testing is out, in its frequentist or Bayesian form, as we saw above. The simple and tedious answer is to make predictions based on whatever premises you have about bent wheels and then verify whether these premises were good. Standard modeling, which of course takes time. What does it mean to suspect the wheel is crooked? That there is the possibility the nature of the cause(s) changed. So much is uncontroversial, but let's see how it fits into the so-called Jeffreys-Lindley paradox, e.g. [183], with an example from Tommaso Dorigo, [61] (using my notation).

A particle counter collects $n = 1,000,000$ instances of some quantum mechanical event, of which $n+ = 498,800$ were "positive" and $n- = 501,200$ were "negative." The details aren't especially interesting to us, except to note that the theory T which informs these counts suggests that numbers of positive and negative hits should be equal. Suppose you fire up the machine and run $n = 1$ instance of the event. Will $n+ = n-$; i.e., will the counts be equal? Obviously not: it is impossible. So is T, that said that $n+ = n-$, wrong? The answer, which may be counter-intuitive is, yes it is. T is false; rather, T has been falsified.

That is, if T says, "In any experiment, $n+ = n-$" and we run an experiment where $n+$ did not equal $n-$, therefore we have falsified T. Not fair? Well, the word *any* inside T does mean *any*: there is no escape. Suppose instead that T actually means, "In any experiment where n is divisible by 2, $n+ = n-$." This is more of a fair playing field. Fire up the detector: $n = 2$ and, say, $n+ = 2, n- = 0$. Is T true or false? False again and for the same reason, $n+$ does not equal $n-$. But wait a second. These are quantum mechanical events, and we expect a T which allows more wriggle room. So suppose T contains premises which allow us to deduce that the probability of $n+$ is $1/2$. From T we can also infer that the probability of $n-$ is also $1/2$. So now if we see $n = 2$ and $n+ = 2, n = 0$, we are no longer sure that T is true or false, because given T these kinds of results can happen. In fact, no matter what n is, if in any experiment we see $n+ = n, n- = 0$—or we see any other values of $n+, n-$—we cannot say that T is false because T says that any sequence of $n+, n-$ can happen. As long as n is observationally less than infinity, which it always will be, no observations can prove T wrong; indeed, since every possible (finite) observation conforms to T, every possible observation confirms T. We can, for fun or for interest, calculate the probability, given T and a fixed n, of seeing some count $n+, n-$.

Enter the "paradox". Dorigo imagines a frequentist statistician measuring $n+ = 498,800$ and $n- = 501,200$. That frequentist, in order to simplify life, calculates $x = n+/n-$ and $s^2 = x(1-x)/n$, and then plugs these values into a normal distribution as the central and spread parameters. The frequentist also accepts that T is true. This lets him calculate the probability that $x < 0.4988$ (the observed fraction) given T is true and that this normal approximation is okay and given the plug-in values for the parameters are uncertainty-free. This calculation gives a probability $p = 0.0082$. Incidentally, the normal approximation isn't really necessary; we can easily do the actual binomial calculation but it gives the same answer to this level of accuracy. So skip worrying about the approximation and worry instead about

what this number means. Well, it is, assuming T is true, the probability of seeing
$n+ = 498{,}800$ or fewer hits in an experiment with $n = 1{,}000{,}000$ runs.

Next Dorigo imagines a Bayesian thinking to himself that T might be true, as he
was told it might be by a physicist. The Bayesian says to himself that "I might as
well suppose that the probability that T is true is 1/2, which means the probability T
is not true is also 1/2. Now T says that the probability of $n+$ is 1/2. But an alternative
to T, call it T', might say that the probability of $n+ = 1/4$. Still another alternative,
T'', might say the probability of $n+ = 2/3$, and so on for other alternatives."

How many alternatives to T are there? If T is the continuum, then uncountably
many. Every number between 0 and 1 (excepting 1/2, which is reserved for T itself)
is a potential alternative. The Bayesian doesn't know which of these uncountably
many alternatives is more likely true than another and so decides to give them all the
same probability: after assuming the probability T is true is 1/2, he spreads out the
other 1/2 over the other possibilities. Then, through the miracle of Bayes's theorem,
the Bayesian can calculate the probability T is true given the observed values of
$n+, n-$, given the assumption that T and its alternatives each had those certain *a
priori* chances to be true. This probability (of T's conditional truth), again using the
normal approximation, is $q = 0.978$.

Here's the so-called paradox. The frequentist "rejects" his "null" hypothesis that
T is true based on the wee p-value, but the Bayesian says it's all but certain that T is
true. Unfortunately, both the frequentist and the Bayesian have produced numbers
which are completely useless and answer no real-life questions. In other words, the
paradox is the gambler's fallacy, done two ways: frequentist and Bayesian.

If T is true, the probability that $n+ < 498{,}800$ is of no interest to anybody unless
they want to be ready for $n+$ values less than this figure (for whatever reason).
Remember, if T is true, any value of $n+$ which is less than or equal to n is possible,
so just because we see one of these values means nothing. Just like the roulette
wheel. The solution to the paradox is that both frequentist and Bayesian were wrong.

Why is it that the Bayesian thought the probability of T being true was 1/2 and
that every other value of probability of $n+, n-$ (implied by different T) was equally
likely? From where did he derive his premises for this bizarre specification? Sounds
like rampant subjectivism. Since the Bayesian started with something absurd,
his result is nothing more than a curiosity. The frequentist based his calculation
assuming either T was true or possibly that it was false, but he offered no alternative
to T, so this is the fallacy of the false dichotomy. Calculating the probability
$n+ < 498{,}900$ given T is true is not calculating the probability T is true. It *assumes*
T is true. If our question is to ask, "What is the probability T is true?" we must
provide alternatives based on justifiable premises or we must just accept that T is
true, just like the roulette wheel.

What we need if we doubt T are realistic alternatives, just like with the roulette
wheel. What are these alternatives? I certainly don't know; at least, not for this
experiment. The physicist might be able to provide them, and even be able to
say (given some other theory to provide a basis) what probability each of these
alternatives is true. If he can, then the Bayesian can work his magic and incorporate
them into Bayes's formula and produce a quantification that, given the evidence of

the alternatives and the experimental data, the probability T is true. If the physicist can't quantify these alternatives—and chances are he can't, since there are too many ways for an experiment to go wrong—he would be better going by his gut. Are there any cables loose? Somebody forget to divide by 2? Probably T + E is true, which is T plus some measurement error. Or perhaps the theory that gave rise to T needs to be altered? Or T really is false, but something like T is true. Do changes to this theory give us models (different T', T'') which better predict the observed data? This is all hard work—unavoidable hard work. Conclusion: there is no paradox, only unrealistic assumptions.

An analysis of the uses and (largely) misuses of the precautionary principle could fill a book, but a word is appropriate here, since the PP relies on uncertainty. There is as always a proposition of interest, say, Y = "The destruction of the word", which is interpreted in some given (horrible) sense. There are a set of premise which are thought to be probative of this "event." "Over-population" was one such set, another is "global warming", and there are many others. A good one is X = "Alien invasion." If hostile aliens from outer-space (or even inner-space, for that matter) were to invade we can assume the worst for humanity. That means, $Pr(Y|X)$ is certain, or is as certain as you like. Now, given X, i.e. given hostile aliens do indeed invade, the costs, whether in dollars, utiles, or whatever, would be incalculable; the highest possible. Therefore, according to the PP, *anything* we do, short of the destruction the aliens would wreak, to protect against this threat will be worth it. And this is so—*if* X occurs.

How many are therefore ready to march on government to demand they protect us from certain alien-induced doom? The problem, as is clear, is that even if $Pr(Y|X) = 1$, there is absolutely zero evidence X itself will happen. Or rather, there are many premises we could accept that would make the probability of X relative to these premises as high as you like. But what we want are premises that are observationally true, that all decisions makers agree on. And there aren't any; at least, not for alien invasion. The problem with the PP is that it is always used to infer that because $Pr(Y|X) = 1$ thus $Pr(X|E)$ is high for some observational premises E which all decision makers agree upon. This is a blantant, but often convincing, fallacy. It is convincing because who wants to be against protecting the world? Another problem is that there are *any number* of X that make Y certain, or nearly certain. You could go on endlessly finding ways to "destroy the planet." According to users of the PP, anything we do to protect against all of them is not out of the question. But this is silly. No, what is needed is to find E that all agree upon which that $Pr(X|E)$ is sufficiently high. If these E can't be found *and* agreed upon by decision makers, then we're left arguing over what's true. As usual.

The two-envelope paradox, also called by Sandy Zabell the exchange paradox, is another good but complicated illustration of the union of probability and decision, see e.g. [45]. (This is a digression that may be skipped.) Before you are two envelopes, A and B. One of them contains X and the other $2X$ dollars, yen, or whatever; some unit of money. You pick one envelope and are (1) asked if you would like to keep it or switch, or (2) open it, view its contents, and then asked if you would like to keep it or switch. Which strategy, keeping or switching, is likeliest to win the big bucks? First the no-peek solution.

The traditional paradoxical solution to (1) is to argue this way. Suppose you pick A, which can be said or assumed to have X. There is then a 50 % chance that B contains $2X$ and a 50 % chance B contains $X/2$. The "expected value" of B is said to be

$$0.5 \times 2X + 0.5 \times X/2 = 5/4X.$$

So clearly you should switch, since the expected value of the envelope you did not pick is larger than X, which is y assumption the value of the envelope you hold. But wait: you could have picked up B, in which case the expected value of A would be $5/4X$, too. (Incidentally, since A can argued to have X or $2X$, each equally likely, the expected value is $3X/2$, but on that reasoning the expected value of B is also $3X/2$.) Of course, there is no possibility of seeing $5/4X$, which is thus never "expected" in the plain English meaning of the word. The phrase has a technical meaning, of course, but that only means the $5/4X$ is some result to a function, and it, too, can never be seen. In some areas of decision theory the goal is to maximize this expected value, which here cannot be done, as either choice gives the same value. And that's the solution. The expected value, which is fictional, cannot be compared to actual values. In other words, it is of no interest that the expected value of (say) B is larger than the supposed actual value of A. True, they seem to both have the same monetary units, but it's still comparing apples and oranges. What's really being compared are, say, expected-dollars to real-dollars, and nobody knows the exchange rate. Equivocation on *expected* has done us in.

This is not likely very controversial, and all it means is that one has to be careful with the expected value criterion of decision making. Ensure only expected or real values are compared, and so forth, or pick another criterion. (Recall here I am not specifically interested in what makes a good criterion.) Now the peeking solution. A is opened and, lo, X is seen. You thus have incontrovertible proof that A has X. Should you switch?

B can still contain $2X$ or $X/2$, right? Not necessarily, if X is real, then maybe that's so, but if X is actual-real, meaning it is discrete and finite, then the situation is different. Suppose the units of X are dollars, as may be reasonably inferred from the game's premises: it makes no difference what currency is used; the only point is that X comes to us in discrete, indivisible chunks (for instance, we could do pennies for the most basic unit). It's also true, we also infer from the game's premises, that if somebody were to really play this game, they would not have an infinite amount of money available. There would be at most N dollars, where N can be as large as you like, just not infinite, so this is no limitation. Suppose you pick A and find it contains \$1. Do you switch or keep? Switch! Because there is conditional on this a 0 % chance B contains \$1/2, which is now an impossible amount because, don't forget, X comes in discrete units and cannot be found in fractions. Indeed, if X is any odd value of X then you should switch, because, given our inferred premises, we are then certain the other envelope contains $2X$! Switching will always double your money if X is odd.

What about even X even? Well, then we are closer to the no-peeking solution, because we might think there is a 50 % chance that the envelope you did not pick has $2X$ and another 50 % chance that it has $X/2$. But if we have more information about N, we can do more. The total amount of money in the game is $2X + X = 3X$, which must be less than or equal to N. Suppose you open A and find some amount W such that $W + 2W > N$. Then you are 100 % sure that B has $W/2$. You should keep, and not switch, any envelope where three times its amount is greater than N (because we know that $W + W/2 \leq N$. We can call these keeper Ws "large." One consequence is that large Ws will always be even. But we should switch when X is even when we know that $3X < N$, because we are gaining information by knowing this fact. Thus, when N is known, and X is real money, we have a solution guaranteed to maximize your profit. To emphasize: this works only when N is known, yet there is nothing in the premises which gives us any information about N.

Of course, we can always add information about N, but that's cheating. So let's cheat: it then turns out that if you know N, then the optimal solution is always to switch. Let's see why this is the case. First assume we know N; we open an envelope and see X. What do we know about X? If it's odd, we switch. If it's even its only possible values are 2, 4, 6,…, up to $N/3$; actually, the nearest integer less than or equal to $N/3$, or floor($N/3$). Why? Well, suppose X is larger than $N/3$, then 3 times that number would be larger than N, which is impossible because we already decided to keep such X. The X we see is, by our tacit assumption, equally likely to be any even number up to floor($N/3$). But the number of the other envelope is not equally likely to be any even number. Why?

Suppose the X we see is 2, then the unopened envelope might be 1 or it might be 4. Thus it is possible that the unopened envelope is odd. Change gears and think about the amount you win in this situation with even X, which is either an even X less than or equal to floor($N/3$) or the value in the unopened envelope. We have just learned that that amount might be odd (if we peeked and saw that X was odd, we would have already switched).

Suppose the unopened envelope is odd. Then X can only be even and only be less than or equal to floor($N/3$). This means that the unopened envelope can only be odd values up to floor($N/3$)/2 (actually, the nearest odd less than or equal to this): it may not take odd values larger than this because that would make $3X > N$. Next, turn this around and suppose the unopened envelope turned out to be 4, then X could have been 2 or 8. But if the unopened value was 6, then X can only be 12 but not 3; if it was 3, we would have already switched. This is the crucial point.

We have already discovered that X is equally likely to be any even number up to floor($N/3$). And we know that the unopened envelope can be any odd value up to floor($N/3$)/2, or it may also take each even value up to this level. But after that, it may only take even values that are divisible by 4. Why? Because only numbers divisible by 4 when divided by 2 are even: and we know that X must be even (and less than or equal to floor($N/3$)).

Let's collect the possibilities of the unopened envelope: (a) odd numbers up to floor($N/3$)/2, (b) even numbers up to floor($N/3$)/2, (c) even numbers divisible by 4 larger than floor($N/3$)/2 up to a maximum 2floor($N/3$). Each of these possibilities are not equally likely: even numbers divisible by 4 but still less than $\text{floor}(N/3)/2$ can appear in two ways (when X is twice or half the value). Odd numbers (a) are just as possible as large even (c), but each (b) is twice as likely. What it means is that the values for the unopened envelope are not equally likely. This is a consequence of two things: the tacit assumption of equally likely game amounts, and our knowledge that X must be even and less than or equal to $\text{floor}(N/3)$. Now, the probability that the unopened envelope contains values larger than $\text{floor}(N/3)$ is non-zero. But X is stuck below that value. Switching can lead to higher win amounts, keeping guarantees lower amounts. So switching makes sense if you'd like the chance of higher amounts. This is a decision strategy that you might not enjoy, of course, but if you did, switching is the way to go. Instead, you might want to minimize risk, or whatever, and decide to stay. But whatever strategy is used relies on these same probability arguments.

Finally, this works when we known N, but since the calculations are the same whatever N is, we don't actually need to know it. So always switching is the best strategy (with the given decision criterion).

Chapter 10
Modelling Goals, Strategies, and Mistakes

A genuine expert can always foretell a thing that is 500 years away easier than he can a thing that's only 500 seconds off.—Mark Twain

An entire book could be written of various implementations of models in the predictive, observable form $\Pr(Y \in y|X,D,M)$ (see the previous chapter for the explanation of this form). Here I can do no more than cover those areas that seem most important to decisions common in science. I emphasize not so much particular models by specific persons, but *how* model results should be communicated and the errors usual in the classical methods. Universally, statistical results are presented as if they were *not* conditional on a model, which of course all are. Over-certainty abounds.

Regression is of paramount importance. The horrors to thought and clear reasoning committed in its name are legion, a fact which is well known, e.g. [56, 83]. But it's more than bad regression: misunderstandings of the nature of evidence are everywhere, but that this is so is increasingly gaining attention; see among many [163, 227], and in the hot field of neuroscience [189]. It's bad enough in academia, but if any reader has experienced consulting in non-academic settings, in, for instance, marketing, you will realize the problems detailed below are trivial. From my many experiences I have been able to discover that ordinary people think statistics is something akin to magic. The discussion on how statistical "control" is not control in the section on regression should be read by everybody.

Reification is *the* deadly sin of modelling. The model is *not* the territory, though this fictional land is unfortunately where many choose to live. When the data do not match a theory, it is often the data that is blamed for marring a beautiful model. Models should never take the place of actual data, though they often do, particularly in regression and time series. Risk is nearly always exaggerated. The fallacious belief that we can quantify the unquantifiable, especially human emotions, is responsible for scientism. Hayek [109], in his Nobel prize speech, cautioned against assuming that the data we have, which is often times the only data we have, must therefore, because of its availability, be causal. This is a form of availability fallacy. Incidentally, Hayek also recommended (a version of) the predictive approach, especially with economic data. "Smoothed" data is often given pride of place over actual observations. Over-certainty is, as I have already claimed, at pandemic levels.

The general, overarching admonition is to escape the Cult of the Parameter. Speak of observables and not parameters. Models should be used in the predictive sense and checked against reality.

Because this chapter describes the some of the many (infinite?) ways probabilistic thinking can go awry, it is more conversational in tone. Finally, at the end, I express some hope about the future.

10.1   The Goal of Models

For those who are cheating and starting at the end, or for those in want of a review, these brief comments on the goal of modeling. The overarching goal of all models is to understand cause. What *caused* the contingent proposition of interest Y to be true? Was it X? Something else? Knowledge of cause is provided by the comprehension of essence, power, and nature; it is the kind of knowledge which moves from the particular to the general. Knowledge of cause is thus always provided through some form of induction, which is why machines can never discover cause. Machines do not have the power to make inductive leaps, no matter how cleverly they are programmed; machines do not have intellects and cannot, however fleetingly or minutely, grasp the infinite as beings with intellects can. Knowledge of cause is never complete and rarely full. We can understand that it is gravity that is causing apples to fall without having total comprehension that space is being warped, though if we do know that, and much more besides, we can make better predictions over a wider variety of circumstance. Knowledge of cause, even though it is apparently closed off to us in its primary mechanism and in the very small, should be restored and understood by all as the primary purpose of science.

Second to understanding of cause, and often an apt substitute, is knowledge of determination. If we know the cause of Y, we also know what determines it. But knowing only what determines Y does not imply we also understand the cause of Y. Instrumentalist accounts of science are deterministic, in the sense I am using the word, as in ascertains. We can have an equation which, fed by X, determines uniquely and precisely the status of Y, but this does not mean X is a cause of Y. Given Y is contingent and because of under-determination, a poor phrase, it could be that what we thought was a cause of Y was not. Under-determination says that, in essence, the contingent always has another explanation, where by *explanation* I mean *determination*. Y only has one cause; it's our knowledge of that cause that is open to doubt. Once again, we meet the distinction between ontology and epistemology. Instead of *under-determination*, perhaps a suitable replacement is just *contingency*. The danger is that explanation or determination will be mistaken for cause, somebody might think Y has more than one cause. Of course, most contingent, or observable, Y do have many forces, i.e. causes, acting in concert to being Y about, but that is only because Y is not simple.

Take Y = "The atmospheric temperature is y". This is not a simple proposition because atmospheric temperature is the product of a vast number of constituents acting in concert, both in the air and at the air-measurement-apparatus junction. There are a plethora of causes of Y. For some measurements, in precisely controlled experiments of isolated (a relative term) environments, these causes may be sought. But the finer and more isolated these experiments grow, the greater the singularity, so to speak, of Y disappears and we must speak not only of Y but of the measurement *and* Y simultaneously. Cause can be influenced by measurement—and not only at the very small. How does an anthropologist measure behavior without influencing it? Answer: he cannot. For most measurements it is hopeless to identify all its causes. Instead, a deterministic approach is taken. With temperature, physicists speak of "ensembles" of gases or of "statistics"; meteorologists might speak of "parcels" or convection. These is a fine approaches, as long as it is remembered, which is often is not, that these terms are deterministic and not causal approaches. They are conglomerations of cause. Knowledge or understanding of cause can be and is had by studying these conglomerations, but the danger of over-certainty is ever present.

The danger is never avoided with probability models, the third level of modeling. Almost universally, users of probability models believe they have identified causes. This, as examples below will prove, is the cause (another cause!) of pandemic over-certainty. Causal models are deductions, and so there is no uncertainty, unless outside events add causes which were unplanned for. Causal models speak ontologically. Deterministic models also have no uncertainty. They say things will happen with certainty. Deterministic models speak epistemologically, though, and not ontologically. Outside causes intrude in deterministic models, too. We have an deterministic equation which says the projectile will certainly be at y at time t. But the location is measured as $y + \epsilon$. This model is formally, strictly falsified since something it said was impossible happened. But everybody treats all deterministic models, and most causal ones, even those models for the very small, in for example quantum mechanics, in probabilistic terms. Look at any argument showing why the results from some collider confirms a deterministic (and probably not causal) projection, even though the observations do not lie precisely exactly with-no-deviation on the curve theory predicted. The departures from the curve do not have to be probabilistically quantified, though sometimes they are. Not all probability is quantifiable. Of course, it can only be, as I said, that if the observations do not lie *precisely* on the theory-determined curve, outside causes *must* have intruded *if* the theory is true. These outside causes are shrugged off even though they provide strict falsification, if we are asking whether and not assuming that the theory is true. The shrugging off is acknowledgement that *complete* understanding of *all* causes in any situation will always be lacking. Understand that the probability spoken of in these shruggings off is of the measurement and not the theory, which is assumed true.

Now probability models say nothing about cause or determination, but an unfortunate and curious culture has developed that associates parameters in probability with cause. Causal hypotheses, or theories, are accepted or rejected depending on the value estimates (or functions of these estimates) of these non-observable parameters take. Even if this made any sense, which I proved earlier does not since these tests are based on the fallacy of the false dichotomy, it cannot be that the certainty one has based on parameters could be as strong as the certainty one has in actual observables. This statement applies to judgements made of "effect size" and the like, too, which also speak of unobservable parameters. It is never remembered that one can be as certain as one likes in the value of a parameter, while remaining mostly ignorant about the observable itself. The certainty (or rather uncertainty) in parameters is always taken as certainty in the observables. Some consequences of this mistake are detailed below. But note that this is only *some*. As I said above, an entire book can be written on the abuses generated by parameter-based statistical methods. For instance, I do not here criticize the methods of so-called structural equations, factor analysis, machine learning, so-called neural nets, and many, many others. But once the material here is absorbed, it doesn't take much work to identify their flaws.

We're sick of hearing about this, but these probability models do not identify cause. They may help understand cause, in a weak way, and they lend some interest to comprehending determination, but their goal in life is only to say how our uncertainty in Y is related to some premises X. And that's it. This is *all* they do. Especially when the "Ys" are related to human behavior, probability models are mute on cause. Just what are *all* the causes of a person receiving, say, an income of *y* dollars a year? Can you even imagine a successful argument which defends such a list? No. Probability models examine observables like physicians used to examine patients before the advent of surgery. Only the vaguest perceptions as to what is really happening can be had. It is therefore grieve-worthy that all who use probability models speak with such monstrous confidence. It is all a bluff, though. Bluffs that ought to be called. How?

I promised at the beginning a discussion of the We-Must-Do-Something fallacy, a malignant tumor in the body of science. Probability models, just like their causal and deterministic brethren, can and must be verified. That means only one thing: making predictions of observables *never before seen*. I mean *never* as in *never*. Known observables, in concert with facts already known or assumed, are used to posit a model. Since any known set of observables can be predicted with absolute certainty by the trick of arranging clever *post hoc* premises, we cannot trust any measure of how close a model "fits" or explains what is already known. These measures have some slight utility in judging between potential models, yes, but as guides to the future (rather, to the unknown) they are very nearly useless. Instead, models should be posited and then predictions made. If the model makes useful predictions—where usefulness is related to the decisions one makes with a model—the model is good, else it is bad. As I said earlier, this technique is slower than the old way of hypothesis testing, parameter estimation, and the like, which are methods fecund in "results." And it is results which are desired by the We-Must-Do-Something fallacy. This is a fallacy that applies to the generation of results as well as the decisions made by or on the results themselves. How easy it is to say "X *causes* Y because of this wee p-value"! How very productive is the research which flows! No. It's past time to slow down. What was the old saying? Trust but verify. That is how all research, particularly that on human behavior should be conducted.

Given that admonition, which I wish I could make stronger because I am sure that many who are reading it who agree that the *other* guy often makes these mistakes, but he won't see the plank in his own eye, there are some general comments to made about "model building" which I haven't elsewhere spoken of. Suppose a set of premises includes the explanatory (*not* causal) premise sex; call this set of premises X. From that, we can form $Pr(Y|X)$. This is one model. But then, because of internal measures of "fit", the researcher decides to drop sex from the premises. Call this reduced set X_{-s}. Speaking correctly, we now have *two* different models. *Any* change to a set of premises (such that the change produces a new set where the old set cannot be deduced from the new reduced set) produces a new model. Often in science the model will retain its old name, as a convenience. But it is a new and different model just the same. This name-retention makes it difficult to know what is being assessed. Is it the old model or the new one with the same name? But that is merely bookkeeping. Since probability is not decision, and it may be unclear what decisions are to be made on the models X and X_{-s}, what is the researcher to do? Well, if he is not the decision maker, he simply makes two predictions, one conditional on X and one conditional on X_{-s}. We then sit back and wait for reality to decide (conditional on the decision to be made) which is better. Whatever we do, we must never ever never decide on a science-wide criterion of relevance. Relevance depends on decision, and that can never be anything but individualistic.

In the language used here a theory is a model is a hypothesis is a supposition. All predictions, since they are *always* conditional, are scenarios are forecasts. And so forth. Now there comes a time when a theory is either believed or beloved but it hasn't yet met the test of reality, which is the comparison with theory predictions with observables, as outlined in the previous chapter. This is natural. Theories are to their creators like the statue was to Pygmalion: beautiful objects of love, as stated before. It is thus perfectly understandable that theory creators defend their creations. But they cannot do it with probability. Here is a cautionary tale of one attempt. Though it applies to the so-called multiverse, everything said below applies to theories or models of any kind.

There is in physics whether the so-called multiverse exists. It is thought to by some who support a certain theory from which the multiverse is deduced; about the multiverse, see [149]. One physicist, Joseph Polchinski, wanted to prove the existence of the multiverse using probability. From his notable paper, [169]:

> To conclude this section, I will make a quasi-Bayesian estimate of the likelihood that there is a multiverse. To establish a prior, I note that a multiverse is easy to make: it requires quantum mechanics and general relativity, and it requires that the building blocks of spacetime can exist in many metastable states. We do not know if this last is true. It is true for the building blocks of ordinary matter, and it seems to be a natural corollary to getting physics from geometry. So I will start with a prior of 50%. I will first update

this with the fact that the observed cosmological constant is small. Now, if I consider only known theories, this pushes the odds of a multiverse close to 100%. But I have to allow for the possibility that the correct theory is still undiscovered, so I will be conservative and reduce the no-multiverse probability by a factor of two, to 25%. The second update is that the vacuum energy is nonzero. By the same (conservative) logic, I reduce the no-multiverse probability to 12%. The final update is the fact that our outstanding candidate for a theory of quantum gravity, string theory, most likely predicts a multiverse. But again I will be conservative and take only a factor of two. So this is my estimate for the likelihood that the multiverse exists: 94%.

Without taking any opinion on the existence of the multiverse, let the theory, i.e. the very complex set of premises, which include a vast array of metaphysical, physical, and mathematical propositions, from which we can deduce the multiverse be called T. T is a complex proposition, and we are interested in whether T itself is true. Why? Because we know the multiverse is true if T is: the multiverse is a deduction or theorem of T. Polchinski wants to bring in Bayesian theory to answer whether T is true. That was mistake number one.

Mistake two is this statement, "I will start with a prior of 50%." This makes no sense. Theories do not have probabilities. And since theories are nothing but (complex) propositions, neither do propositions have probabilities. Indeed, no thing *has* a probability. Probabilities are measures of knowledge, therefore they have to come equipped with gauges, i.e. conditions. In other words, all probability is conditional, as earlier proved.

Many think one natural gauge is the proposition W = "T might be true", which is logically equivalent to "T is true or it is false". Both of these are tautologies, which we know are true conditional on our knowledge of logic and understanding of English grammar. But it makes no sense to say, as Polchinski said, $Pr(T|W) = 50\%$. Tautologies are non-informative. The best we can do, as I pointed out earlier, is to deduce T's contingency, which gives it a unit interval probability, *sans* endpoints. Of course, Polchinski may not have had the tautology in mind, but some other gauge. Call this G, which relates to come complex proposition in Polchinski's head. Then it might be true that $Pr(T|G) = 50\%$.

But what would this G have to look like? Well, it would have to be directly probative of T itself, which means of the propositions of which T is composed. And if Polchinski really had such a G, it is more plausible these G-propositions would already be in T to give it support. Why withhold from T knowledge relevant to multiverses? It doesn't make sense. But then G might have nothing probative to say about T except its contingency like W, in which case $0 < Pr(T|G) < 1$.

According to the rules of probability, $Pr(T \text{ false}|G) = 1 - Pr(T|G)$. But what does it mean to say T is false? Just that at least one of propositions within T is false. And if we knew *that*, then we would never entertain T. We would instead modify T (which really means making a brand new T) to remove or transform these troublesome propositions. If G told us which part of T was wrong we would fix it. Put all this another way. If all we had in contention for the multiverse was T, then T is *all* we have. We can't judge its truth or falsehood because we have nothing to compare it to. T is it. It's T or bust.

I'm sure (though I didn't check) Polchinski's numerical calculations are on the money, but the end result is meaningless. T has to be compared not against some internal gut reaction, because there is no such thing as subjective probability, but against the predictions T makes or against rival theories, which provide the only natural comparators. That is, Polchinski might have some alternative theory M in mind, a rival to T such that, given M, the multiverse is not a theory of M. Polchinski's G makes a little more sense.

There may be, and almost certainly are, overlapping elements of T and M, sub-propositions which they share. Nevertheless, T is not deducible from M, nor vice versa, else they would be the same theory. We've already seen that it makes little sense to have in G propositions which duplicate the multiverse-predictability propositions in T, and the same objection applies to M. That means G is something else. The simplest would be the "freshman" G, which is "There are two rival theories, T or M, and only one of which can be true" (and is the only thing a freshman in physics is capable of knowing). Therefore, $Pr(T|G) = Pr(M|G) = 1/2$, via the statistical syllogism. But that's as far as we can go without additional evidence, such as observations of the multiverse (which won't be had, it is claimed) or via observables deducible from T or M. Other G are possible, but it is easy enough to see, since there is no such things as subjective probability, we're up against the unquantifiable. Gut feeling as a decision takes the place of probability. In other words, it's better to go out and find proof of T or M.

Naturally, everything said holds for theories of any kind, as I said. Readers will also see in the criticism of Polchinski the standard argument why hypothesis testing, whether by wee p-value or Bayes's rule, is based on the fallacy of the false dichotomy.

10.2 Regression

We have two observations and want to discover if there is a relationship between them, in the sense that knowledge of one is relevant to knowledge of the other. The observations take the form of propositions, e.g. Y = "The value of object one is y_1 and X = "The value of object two is x_4" (these are labels and not measurement numbers). In classical language, we can apply "correlation" or "regression." Those techniques are rather restrictive, however, as they specify "straight-line" relationships between the parameters of the observations, which can only be had by assuming continuity. If the two observations were unrelated to one another, then knowledge of one would be irrelevant everywhere to the other. If at some level of the X the probability Y takes any of its values differs from the same probabilities for at least one other level of X, the two observations are related, or "correlated", as it were.

Before us are p bins, mapping to the levels in X; that is, the possible values for the X proposition are $(x_1, x_2, \ldots, x_p)$. In each bin there are N objects which can take

m states, corresponding to the possibilities of Y; that is, the possible values for the Y propositions are $(y_1, y_2, \ldots, y_m)$. Another way to say it is that each bin or level of x_1 contains objects with the labels $y_1, y_2, \ldots, y_m$. And those are all the premises we need for discovering some functional probability relationship between X and Y— but not necessarily a linear one. Indeed, a strictly linear function will be rare and difficult to realize since X and Y form a $p \times m$ grid.

Once we observe actual values, no statistical model is needed to display what happened. The data can be plotted after, as is hoped, compressing to a smaller grid comprised of decisionable, i.e. measurable, points (in other words, make p and m the minimum necessary for making decisions). The standard x-y plot is, after all, such a grid, with points perhaps scaled to the number of overlapping observations per grid point.

The premises as we have them say nothing about the relationship between neighbor (near or far) X bins to any particular y_i; meaning knowledge in what might happen when $X = x_i$ is irrelevant to knowing what might happen when $X = x_j$. High values of Y might be more likely at $X = x_i$, and low more likely at $X = x_{i+1}$, and again high more likely at $X = x_{i+2}$, and so on. If we desire there to be a relationship, like something resembling a line, we have to specify it in the premises.

What might these premises look like? That depends on the application and information at hand. For example, classical regression assumes that higher (or lower) values of Y are likely for higher values of X. The nature of this increase (or decrease) is parameterized, a situation we want to avoid because parameters skate over evidence, or rather, they introduce complex premises into situations which are often unwarranted.

Our goal, as ever, is to discover, for as-yet unknown observations, $\Pr(Y = y | X = x, D, M)$. This can be plotted in the same manner as the scatter-plot. It will be a $p \times m$ grid with probabilities at each point, specifying uncertainty in not-yet-known values of Y. These can be displayed, for instance, by variously sized plot points or by shading each grid point, like heat maps. Or simple tables can be created showing only those probabilities which exceed some threshold important for decisions, as we did in the last chapter. Extreme caution should be taken to summarize *all* available information. For instance, tables showing the most likely value of Y for each level of X, or the "expected" value of Y, will be tempting because they are simple, but much information is lost in this maneuver and over-certainty encouraged.

If X is irrelevant to Y, then for each level of X, the probability of new Y taking each of its values will be equal. Of course, any two sets of numbers pulled out of thin air, given these premises, are likely to exhibit relevance at some levels of X. The departure from irrelevance thought important thus not unexpectedly depends on the decisions to be made.

Most modeling errors are found in regression simply because regression is used more than any other technique. Like all models assuming continuity, regression models are falsified in fact because they assign zero probability to events which have been or will be observed. So if they are to have any use at all as approximations, it is required they give reasonable (where they term is flexible and defined by decisions based on the model) probabilities to intervals and not points of interest.

Regression models often suffer from undiagnosed probability leakage. This happens when Y constrained but the continuity isn't. For example, income is usually constrained to be greater than 0, grade point averages are between 0 and 4, answers to *ad hoc* survey "instruments" (see below) are often limited to take one of only several discrete positive values, and so on. Of course, all real observables are constrained in this sense because none can be infinite or continuous (see the discussion on measurement in Chap. 9). But some are more constrained than others.

Classically, a regression begins when the *ad hoc* assumption (premise) that the uncertainty in some proposition Y can be quantified using a normal. Normals have two parameters, a central *m* and a spread *s*. A premise is added that the central parameter is a (often linear) function of explanatory, probative X variables. *Pace*

$$m = \beta_0 + \beta_1 x_1 + \beta_2 x_2 + \cdots + \beta_p x_p, \qquad (10.1)$$

where in frequentist statistics no information is given about the β or s, and where in Bayesian statistics "flat" or other priors express uncertainty in these parameters. Like all parameters, these do not exist; they can never be observed nor verified. Reification happens here (and *all* the time) when this equation is thought to be of the observables themselves, and not the unobservable parameters.

The goal of regression, almost never stated and realized as often as come honest answers from political party spokesmen, is to make statements of the kind "Given our modelling assumptions and the data we have observed, the probability of Y is p when $X = x$", where the Y and X are propositions of the sort with which we are now familiar. The thing to note is that Y will be a proposition about an interval, such as Y = "The value of y will be between y_0 and y_1". Single values of y are impossible in continuous models, but single values of x are usual, e.g. X = "The value of $x_1 = w_1 \ldots$ and the value of $x_p = w_p$". This is eminently commonsensical, and just what the customer wants. But it is not what is done. Instead, classical procedure revolves, as it does in most methods, around making statements about the β, and, sometimes but rarely, about s.

Civilians do not care about the parameters, but they are all the statistician will talk about. And it is because of this reason that civilians, and then statisticians because of their interaction with civilians, forget what parameters are and reify them into observables and, worse, into believing parameters are proof of efficient causality. The latter is the well known fallacy of supposing causation follows from correlation, and the supposing of correlation comes from inflating the importance of parameters, which is the Deadly Sin of Reification.

I don't want to single out any individual, but headlines like the following are typical "Restaurant rage: Living in an area with lots of fast food stores can make you impatient and unable to savor things, researchers warn." A reporter fashioned this after reading a peer-reviewed paper, or, more likely, a press release about the paper [114]. What did that paper say?

Researchers recruited people on-line, asked them in which zip code they lived, looked up in a book how many fast food restaurants existed in that zip code, ignoring that land area and land use in zip codes can change dramatically, asked the participants some questions, quantified the answers to these questions, and ran a regression which showed the answers to the questions and number of restaurants (actually, the ratio of fast food to "normal" restaurants) were "correlated" in the sense that parameters in the regression evinced wee p-values. The researchers then committed the Deadly Sin of Reification and announced that their work *showed* "that as pervasive symbols of impatience, fast food can inhibit savoring, producing negative consequences for how we experience pleasurable events."

Not to be too unkind, but this is nuts. But *exceedingly* typical. I also don't want to exaggerate, but *entire fields* make their living abusing statistics in this way. As a sort of hobby, I have been collecting particularly egregious examples of such abuse.[1] There is some harmless fun to be had exposing the errors committed, but no matter how often or how many times I've done this, the assumption is always that it is the studies I have found which are flawed and that others are fine. Maybe they're not perfect, but they're close enough.

This is false. Entire areas of sociology, education, psychology, really any of the so-called soft sciences, are beset with "results" given by regressions that, if not outrightly false, are believed with a certainty far, far exceeding what is justified. Indeed, it is more likely than not that any regression you have seen is in error. And if you yourself have created a regression, you have very probably contributed to the harmful pandemic of over-certainty which besets science and which has given rise to scientism.

An entire book could be written on the abuses of knowledge done with regressions, but this isn't it. But you don't have to take my word for it (see e.g. [27]). Given here are the warning signs which can be used to check the next regression you come across.

We need an example. Suppose a researcher invents a 1–7 "hate score", with higher numbers indicating more hate. About these kinds of maneuvers, more below. The researcher "applies" his "instrument" to a group of volunteers, probably his students, simultaneously measuring other characteristics of this group, such as each volunteer's sex, age, and so on, and perhaps also their answers on other "instruments" popular in that researcher's field. The data is collected and a regression run. Incidentally, I've heard people say they "submit" data to the software package, as if that package will authoritatively give the one correct answer. This is a hint of the problems to come.

In this regression model will be a parameter related to sex. If it has a wee p-value, the researcher will make the mistake of saying, "Men and women hate differently." Now they may, but this wee p-value is certainly not proof they do. The researcher will also make the mistake of saying his "hate" score unambiguously and without error *is* hate. Hate is a complex emotion, but the researcher will believe his one-number of summary captures all or most of what is important. Only that which is measurable becomes important. This is scientism.

---

[1] See my web page http://wmbriggs.com.

In his data, a certain number of women answered in a certain way, and so too the men. Their answers can be tabulated. Suppose more women gave higher answers than men. Which is evidence of what? Beside the obvious that these women gave higher answers than these men, it is also positive evidence that other women "like" these in situations "like" this, will give higher answers to this set of questions than other men "like" ours in situations "like" this. And this can be quantified. Indeed, we can form statements like this:

$$\text{Pr(Woman has higher hate score than man}|D,M) \qquad (10.2)$$

where M contains the premises which led to the normal and regression and D the observed data. This is a *predictive* equation. Bayesians would call it the predictive posterior; frequentists the forecast or prediction. There are *no* parameters here: they are "integrated out." We are in the pure Land of Observables only.

The *p*-value for sex was wee, but it could very well be the probability in (10.2) is 0.501. This kind of thing *happens commonly*: try it yourself. I mean, classical answers will show "significance" where the predictive probability difference is miniscule. It is then *true* that, given M and D, it is likely the next woman of the type represented in M will have a higher hate score than the next man of the type presented in M. Consider that if there were no differences in the way women and men of M acted on hate scores, then the probability the next woman scores higher would be 0.5. So is 0.501 a large enough departure from 0.5 to be important? I have no idea. That depends on the decision one would make regarding that difference. Because I can't see how something so trivial can be important doesn't prove it isn't. But the researcher surely has the burden of demonstrating *how* it can be important. He has to provide scenarios relevant to the decision. Or he can agree it is trivial for himself and his audience and he can *drop sex from the model.*

There is more to it. There are other variables besides sex in the model. That 0.501 is the probability of M women answering differently than M men *conditional* on specific values of those other variables. All variables in any model must be specified to compute probabilities like (10.2). Any one configuration, or set of all values, can be called a *scenario*. That 0.501 is for one scenario. It may not hold for others. It might be, for instance, that for some other scenario this probability rises to, say, 0.7 or as low as 0.28. Who knows? Two things must be drawn from this.

The first is a reminder that sex is irrelevant, conditional on M and D, if for all scenarios, probabilities like (10.2) are constant. Of course, (10.2) is not the only possible relevant probability. Another might be, given M and D, the probability of women scoring 7 more often than men; another might have to do with the average hate score, and so forth. These propositions of interest, as I said many times, are not statistical matters. They are driven by the decisions to be made. It must be ever emphasized that these decisions are rarely matters of science (this depends on the field and the uses to which the data are put, of course). And that means it is not incumbent upon the statistician to define pertinent propositions of interest. These must be given by the (as it were) clients, who, experience shows, can be awfully lazy and want decisions to be made for them (this of course does not apply to you, dear reader).

And this leads to the second conclusion. The pertinent scenarios are also for the clients, not the statisticians, to decide, although some rough guidelines can be stated. Now it might have been in the set of data at hand the age range was 18–22, but the pertinent scenario relates to age 24. This can be computed. And since we are calculating predictive probabilities, there is no difficulty calculating probabilities for any value not seen in the data, or for the values seen. Suppose the proposition of interest is (10.2). We have several built-in scenarios with which to calculate (10.2), which are the observations themselves. Thus (10.2) can be calculated for every observation on hand, as if that observation were a new scenario, and the data plotted. To make it interesting, it can be sorted and the points examined for which values of the other variables are associated with low and high values of (10.2). A conclusion might be, say, "*Ceterus parabis*, the probability a woman like those surveyed has a higher hate score than a man like those survey is about 40% for those under 20 years, about 50% for those 20 to 22, and roughly 62% for those 23 and older." Those demarcations and "about"s and "roughly"s are *decisions* made by the client.

It might not have been clear, but model-building strategies are not necessarily difficult using this predictive-observable approach. The variables pertinent to the client or researcher are left in, and those which are not are left out. Simple as that. Well, not so simple, perhaps. The propositions of interest and pertinent scenarios are not always easy to come by. But if they are not known, then why is the researcher conducting an investigation?

Two words of caution. Wee *p*-values do not always lead to practical relevance, nor do non-wee *p*-values always lead to practical irrelevance. As I have said often enough, *p*-values should be abandoned with extreme prejudice. Never look at them again. Look only at the probabilities of pertinent propositions of interest calculated from researcher-drive scenarios. Secondly, I beg the Lord above that we not become fixated on a probability which is considered "good" or "significant". There is no such thing. Goodness or real-life significance depends on the decisions to be made. They are not statistical questions. To think they are is to engage in scientism.

Too many think they can discover the perfect model. This is not impossible in theory, but it would be a full causal or deterministic model and not a probabilistic one. What is forgotten is that all probability is conditional. Therefore the probabilities derived from any model are correct (barring calculation error or cheating). A regression does *not* say what causes what. Better models, as discussed previously, are those that lead to better decisions, and that is measured by verification scores.

A researcher may be interested in the relevance of specific probative propositions. These are entered into the model and probabilities of propositions of interest are then calculated—given scenarios, which include values of other propositions the researcher chose to put in. Practical relevance, as always, depends on the decisions to be made. The researcher can, of course, make these decisions for his audience, but he must justify them. If it is not already obvious, the lesson is this: there is no magic formula or way to discover the best model. The process can be automated, however, as long as all the decisions are made in advance. But since these decisions won't be the same from researcher to researcher, this automation is—and should remain—limited. A researcher should *not*, if he set out to discover differences in the sexes relating to a hate score, say anything like "There are no differences between men and women." Unless strict irrelevance were discovered for all possible scenarios, this will never be true. Instead, it is the researcher's duty to state the probabilities for (10.2) (or whatever) for those scenarios deemed important.

A word about "control". In classical language, secondary "variables" are entered into models like (10.1) and are said to be "controls", e.g. "We controlled for age, sex, and weight". There will be some Y and some main X, as always, and a blizzard of extra information that may or may not be of interest. See [229] for some examples in economics. These extra or secondary propositions are the controls. The word "control" here is spurious, an awful misnomer. Even in controlled experiments, in the real sense of purposeful manipulation, as in say chemical "bench" work, these secondary propositions are *not* controlled. For instance, a drug trial will allocate, i.e. enforce the control, that half the patients eat the drug and half the placebo, but where the secondary characteristics like age, sex, weight and so on are not controlled in the least. Of course, many experiments are more complex and the giving of the drug and placebo is balanced between more than one characteristic, like for instance sex. Yet there will still always be other characteristics that are not purposely controlled, and these are called *controls*, which is the exact opposite of the truth.

The mistake arises because, even when people know better, they cannot help but think (10.1) is a causal or at least deterministic model. (The predictive, logical approach advocated in this book will not eradicate this dogged fallacy.) Suppose our trial—we know we do not need "randomization"—controlled, in the real sense, for group, drug or placebo, and sex. Age and weight are incidental but measured, as are, if the trial is typical, dozens of other things, including (as is now usual) the scores from questionnaires (see below). There was nothing done in the experiment to control the age beyond maybe allowing only adults and even less done about weight. Could age or weight *cause* a change in the actual response? Not in (10.1), which is pure probability and therefore not indicative of cause, but could a patient's weight causally change the outcome? In most experiments, given what we know of medicine, the answer is yes. If this is true for our experiment, the only way to tell is if we truly control weight. As it is, weight is incidental, and whatever it is that is causing Y—and there are many possibilities and perhaps many partial causes—might also be causing weight to vary (in some way).

Once more, (10.1) says *nothing* about Y. It says something about a parameter (and only one of two) used to represent uncertainty in the value of Y. Even if we knew weight is a partial cause, the way it enters into the model is as a function of a parameter and not Y. Thus it still makes no sense to call these secondary propositions controls. What are they then? They are propositions which change our uncertainty in Y, *conditional on* the data we observe, the model and the other propositions we have already entered into the model. And nothing else. When a user of a statistical model says he has "controlled" for certain "variables" he is conveying to his audience, and probably to himself, the false impression that he knows what is causing what. This kind of language is thus harmful even when it used as a shorthand. The temptation to claim more certainty than is warranted is impossible to resist.

Reminder! As discussed in the last chapter, what *should* happen is this. The pertinent questions are decided, the experiment designed, the data collected, the model built, and then, every single time, the model should be used to make predictions. And then *we must wait until those predictions are verified* before pronouncing the model good or bad. This is the *only* way we can assess the value of the model, to see whether it is any good. This is the way it is done in better hard sciences. The model is announced as *part of* the findings, not hidden away and assumed to be proper as it is in nearly all statistical analyses. An excellent summary of what can go wrong with health data and nutritional guidelines is in [7].

10.3   Risk

A common way to report statistical models is by relative risk. The simplest way to think of it is this: the probability of having a disease, or of dying or of having or evincing whatever outcome of interest, given one exposure, divided by the probability of disease given another exposure. Common exposures (for ease) are high and low. A relative risk of 2 means the disease is twice as likely to be found in the high group. Relative risk is a misleading way to report, however, because it exaggerates risk, as I'll prove. It must, too, be kept in the back of your mind that the goal of all scientific research is to discover cause—where possible.

We'll return to the particulate matter example from Chap. 7 on causality, which should be reviewed. Recall we had two groups of 1000 each, one exposed to no PM2.5 (the low group) and one exposed to some high level. Five folks in the low group contracted cancer of the albondigas, and 15 in the high group did.

What's the difference between a probability of one in ten million and one of two in ten million? The official answer to most decision makers is "Not much," but the relative risk is 2, which is considered high. If the two numbers were the (conditional on evidence) probability of disease in the low and high groups, one could honestly write a paper that said, "Exposure to high PM2.5 doubles the risk of cancer." But it would highly misleading.

What counts as "exposed"? Daily inhalation? If so, for how long a period and in what quantities? And what about the people "not exposed (to high PM2.5)"? They were "not exposed" because they were *certainly different* than the people who *were* "exposed." That they must be different is a logical truism: if they were the *same* they would have been exposed to high PM2.5, too. Since the low group were not exposed, they are different. In how many ways are the people in the two groups different? Nobody knows, and nobody *can* know. The number of measurable differences, as we already saw, is astronomic.

Is our job made easy by defining "exposed to" as "any exposure of any kind in any high amount", which is certainly plain enough? No, because it begs the question why the folks who did not see any exposure of any kind in any amount didn't. Were they off on holiday? Did they eat different foods? Come from a different culture? Were they older or younger? Have variant genetics? You can go on and spend a lifetime and never be sure of identifying all things different between the groups.

Those problems are too hard, so let's do something as easy as possible, which will still lead to an important twist. Suppose God Himself told us that the probability of cancer in the exposed (in any way to high PM2.5) group is 2 in ten million and the probability of cancer in the not-exposed group is 1 in ten million. The relative risk is 2.

Evidence suggests that Los Angeles is a "hot spot" for high PM2.5 exposure. Now L.A. proper has about four million residents, some of whom will have been exposed, some not. Suppose for the sake of argument the population is split: half are exposed, half not. Remember: the Lord Himself assured us the relative risk of 2 is correct. The chance is thus 67 % that *nobody* in the exposed group will develop cancer. Pause here. That's *no*-body. Meaning there's a two-out-of-three probability that not a single soul of the two million exposed people will get cancer. Anti-intuitive? Shouldn't be: even with a relative risk a whopping 2.0, there's still only a paltry 2 in ten million chance of disease for any individual.

Likewise, we can calculate the chance nobody gets cancer in the not-exposed group: it's 82 %. Isn't that 67 % versus 82 % odd? In the exposed group, the chance at least one person gets cancer is one minus the chance nobody does, or 33 %; and the chance at least one person gets it in the non-exposed group is thus 18 %. The risk ratio is now 1.8, a big change from 2! But how can this be when God Himself told us the relative risk is 2? That's because the "2" is an abstract number. When we start applying it groups of real people, the stark difference begins to disappear. For example, if we applied the same calculations to the population of New York (about eight million) the real relative risk is 1.67. Lower still. By the time we consider the population of the entire United States, the real relative risk dwindles to 1.

Why is this happening? Because relative risk is stated in terms of *single people*. When you're concerned only with yourself, it's the right calculation—but a silly one, because you're better off knowing the probabilities involved. Consider that you have the same relative risk moving from 2 in ten million to 1 in ten million as from 1 in 2 to 2 in 2, yet the situations are worlds apart.

Let's think again about the population of LA. The chance that *just one* person of the two million in the exposed group develops cancer is 27 %. That means there's a 94 % that either nobody gets it or just one person in two million does. And there's a 16 % chance just one person in the non-exposed group gets it, or a 98 % chance 0 or 1 people get cancer. The overwhelming probability—94 %—is *no more than one* poor soul develops cancer in the exposed group, and that it's more likely *nobody* does (67 %). The same is true in the not-exposed camp.

Here's another way to look at it. In all of LA's population, some people in each group might get cancer. The chance that *more* people develop cancer in the exposed group than in the non-exposed group is only 28 %. Which is not even close to 100 %. Even with a relative risk of 2, there's only just over a 1 in 4 chance of larger numbers of cancer patients in the exposed group. The largest chance, 59 %, is that both groups will have the *same* number of people who develop cancer. And there is even a 13 %

chance that the not-exposed group will have *more people* who develop cancer than the exposed group! There is an enormous difference between sensational relative and sober absolute risk.

Does 1 in ten million seem low? It doesn't to the EPA. In one of many guides [69, p. 5], they fret over risks as tiny as that. However, they often cite risks of 1 in a million and 1 in ten thousand as regulation-worthy. Let's use the latter and see what boosting the exposed cancer chance to a whopping 2 in ten thousand does to our calculations.

We first need a workable relative risk—2 is really too large. We need one considered by the regulatory establishment high enough to warrant hand-wringing. Use 1.06, the high-water relative risk in a series of widely touted papers by Michael Jerrett and co-authors [125] who measured exposure to dust. Jerrett spoke of many diseases, mainly cardiac and respiratory ailments, but what matters here is the size of relative risk touted as worrisome. We're still claiming that all that counts is exposure or non-exposure; all or nothing. We can't do here the harder problem of multiple levels of exposure and the reasons for differences in exposure, but the sophisticated reader can easily see the extensions. Assume some people had "enough" exposure to dust and others didn't. The disease for the sake of illustration is still cancer, but it could be anything. With a relative risk of 1.06, the chance of disease in the not-exposed group is 0.000189. Figure 10.1 shows a picture of the probabilities for *new* disease cases in the two groups in LA, still considering a population of four million and split exposure.

There's a 99.99 % chance that from about 300 to 440 not-exposed people will develop cancer, with the most likely number (the peak of the dashed line) about 380. And there's a 99.99 % chance that from about 340 to 460 exposed people will

Fig. 10.1 The probability citizens of Los Angeles develop cancer with and without exposure to high PM2.5. These are not normal distributions, but binomial and only appear smooth. These are non-zero probabilities for observable events

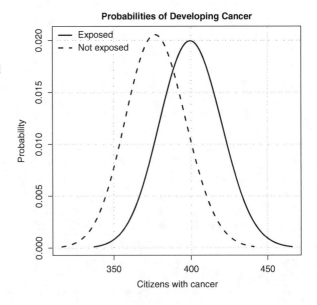

develop cancer, with the most likely number about 400. A difference of about 20 folks. Surprisingly, there's only a 78 % chance that *more* people in the exposed group than in the not-exposed group will develop cancer. That makes a 21 % chance the *not-exposed* group will have as many *or more* diseased bodies.

This not-trick question helps: how many billions would you pay to reduce the exposure of high PM2.5 to zero? How many people would get cancer? The answer is: *this statistical model says nothing about what causes cancer*. As shown in Chap. 7, we have to *assume* PM2.5 is a cause. We *can* say is that eliminating high PM2.5 eliminates high PM2.5, which is equivalent to saying that everybody else, all four million folks, would be exposed to low PM2.5. This is important. Calculations show there's a 99.99 % chance that anywhere from about 650 to 850 people would get cancer, with the most likely number being around 760.

There's a 99.99 % chance that from between just under 700 to just over 860 people will have cancer, given exposure is split in the population between high and low PM2.5. And there's the same chance that from about 670 to 840 people develop cancer assuming nobody is exposed to high PM2.5. The most likely number of victims for the split population is 777, and it's 754 in the all-low population, again, only about 20 folks different. There's only a 71 % chance more people in the split population would have more people with cancer than in the all-low scenario, but there's also a 28 % chance more people in the all-low scenario have cancer.

This and not the previous picture is the right way to compare real risk—when the risk is known with certainty. It's not nearly as frightening as the normal methods of reporting.

Figure 10.2 shows there's a cap, a number which limits the amount of good you can do (regardless whether cancer is caused by PM2.5 or something associated

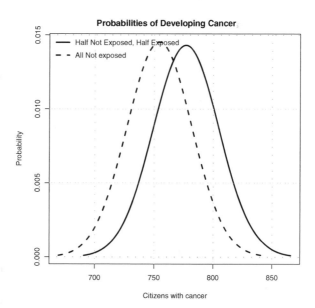

Fig. 10.2 The probability of the number of people having cancer, when half the population of LA is exposed and half not, compared to the supposing the entire population isn't exposed

with it). If we use the 99.99 % threshold (adding a few 9s does not change the fundamental conclusion) and we eliminate *any possibility* of high exposure, then the *best* we could save is about 200 lives—assuming, too, the cancer is fatal. That comes from assuming PM2.5 is cause, that 866 exposed people develop cancer and 670 not-exposed people get it (the right-most and left-most extremes of both curves). There's only a 0.005 % chance that 866 or more exposed people get cancer, and there's a 99.985 % chance at least as many as 670 in the not-exposed people get it. The *most likely* scenario is thus a saving of about 20 lives. Out of four million. Meaning our Herculean efforts to eliminate all traces of dust at *best* we'd affect about 0.004 % of the population, and probably more like 0.0005 %. And never forget we only assumed PM2.5 was a cause. If it isn't, all our efforts are futile. How many billions did you say?

There are other strong assumptions here. The biggest is that there is *no* uncertainty in the probabilities of cancer in the two groups. *No* as in *zero*. Add *any* uncertainty, even a wee bit, and that expected savings in lives goes down. In actual practice there is plenty of uncertainty in the probabilities. The second assumption is that everybody who gets cancer dies. That won't be so; at least, not for most diseases. So we have to temper that "savings" some more.

Assumption number three: exposure is perfectly measured and there is no other contributing factor in the cancer-causing chain different between the two groups. We might "control" for some differences, but recall we'll *never know* whether we measured and controlled for the right things. It could always be that we missed something. But even assuming we didn't, exposure is usually measured with error, as we have seen. In our example, we said this measurement error was zero. In real life, it is not; and don't forget Jerrett relied on the epidemiologist fallacy. Add any error, or account for the fallacy, and the certainty of saving lives necessarily does down more.

Let's add in a layer of typical uncertainty and see what happens. The size of relative risks (1.06) touted by authors like Jerrett get the juices flowing of bureaucrats and activists who see any number north of 1 reason for intervention. Yet in their zeal they ignore evidence which admits things aren't as bad as they appear. Here's proof.

Relative risk estimates are of course produced by statistical models, usually frequentist. That means *p*-values less than the magic number signal "significance". Now (usually arbitrarily chosen and not deduced) statistical models of relative risk have a parameter or parameters associated with that measure. Classical procedure "estimates" the values of these parameters; and as we've seen, the guesses are heavily model and data dependent. Change the model, make new observations, and the guesses change.

There are two main sources of uncertainty (there are many subsidiary). This is key. The first is the guess itself. We thus far assumed there was *no uncertainty* of the first kind. We *knew* the values of the parameters, of the probabilities and risk. God told us! Thus the picture drawn was the effect of uncertainty of the second kind, though at the time we didn't know it. We saw that even though there was zero uncertainty of the first kind, there was still tremendous uncertainty in the

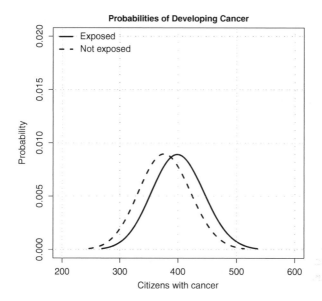

Fig. 10.3 The probability citizens of Los Angeles develop cancer with and without exposure to high PM2.5, factoring in parameter uncertainty

future. Even with "actionable" or "unacceptable" risk, the future was at best fuzzy. Absolute knowledge of risk did not give absolute knowledge of cancer.

This next picture, Fig. 10.3 shows how introducing uncertainty of the first kind— present in every real statistical model—*increases* uncertainty of the second.

The narrow, highly peaked lines are repeated from before in Fig. 10.1, which were the probabilities of new cancer cases between exposed and not-exposed LA residents assuming perfect knowledge of the risk. The wider lines are the same, except I've added in parameter uncertainty (since I don't have Jerrett's numbers this is only a reasonable guess). In Bayesian terms, we "integrate out" the uncertainty in parameters and produce the posterior predictive distributions. The spread in the distribution doubles: *uncertainty increases dramatically*.

There is also more overlap between the two curves. Before, we were 78 % sure there would be more cancer cases in the exposed group. Now there is only a 64 % chance: a substantial reduction. Pause and reflect. Parameter uncertainty increases the chance to 36 % (from 22 %) that any program to eliminate PM2.5 does *nothing*, still assuming PM2.5 is a cause. Either way, the number of affected citizens remains low. Affected by cancer, that is. *Everybody* would be effected by whatever regulations are enacted in the "fight" against PM2.5. And don't forget: any real program cannot eliminate exposure; the practical effect on disease must always be less than ideal. But the calculations focus on the ideal.

As above, here in Fig. 10.4 are the real curves we should examine.

This is astonishing. By properly accounting for uncertainty, there is now only a 58 % chance that more citizens would develop cancer in the spilt-exposure group

Fig. 10.4 The probability of
the number of people having
cancer, when half the
population of LA is exposed
and half not, compared to the
supposing the entire
population isn't exposed,
factoring in parameter
uncertainty

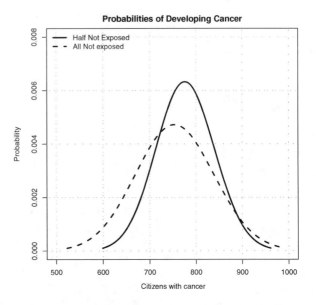

than in the all-not-exposed group. And there is a 41 % chance that more people
would have cancer in the all-not-exposed group. There is more uncertainty in the all-
not-exposed group, too, meaning the number of lives saves is up in the air; almost
unknowable. This is the benefit of examining predictive uncertainty. It gives the
best, of course still model and data dependent, picture of what we can expect.

We're not done. We still have to add the uncertainty in measuring exposure,
which typically is not minor. For example, Jerrett (2013) assumes air pollution
measurements from 2002 effected the health of people in the years 1982–2000. Is
time travel possible? Even then, his "exposure" is a guess from a land-use model,
i.e. a proxy and not a measurement. Meaning he used the epidemiologist fallacy to
supply exposure measurements.

I stress I did not use Jerrett's model—because I don't have it. He didn't publish
it. The example here is only an educated guess of what the results would be under
typical kinds of parameter uncertainty and given risks and exposures. The direction
of uncertainty is certainly correct, however, no matter what his model was.

And there are still sources of uncertainty we didn't incorporate. How good is the
model? (The only "true" model is a fully causal one.) Classical procedure assumes
perfection. But other models are possible. What about "controls"? Age, sex, etc.
Could be important. But controls can fool just as easily as help, as we now know.
Anyway, controls don't change the interpretation. If there were controls, we would
just have different pictures for each subset (say, males versus females, etc.).

All along we have assumed we could eliminate exposure completely. We cannot.
Thus the effect of regulation is always less than touted. How much less depends on
the situation and our ability to predict future behavior and costs. Not so easy.

I could go on and on, adding in other, albeit smaller, layers of uncertainty. All of which push that effectiveness probability closer and closer to 50 %. But enough is enough. You get the idea.

10.4 Epidemiologist Fallacy

One fallacy is so common and so harmful that it deserves special mention. It is the *epidemiologist fallacy*. It is born from the more well-known ecological fallacy, but I prefer the neologism because without this fallacy, most epidemiologists, especially those employed by the government, would be out of a job. The epidemiologist fallacy is also richer than the ecological: it occurs whenever an epidemiologist says, "X causes Y" but where he never measures X *and* where he uses classical statistics to claim proof of a cause—based on, say, wee p-values or large Bayes factors; see [27]. Over-certainty is guaranteed.

How frequent is this fallacy? As rare as a lobbyist on the northward-bound Acela out of Washington D.C. on a Friday night. Over the years I have cataloged hundreds of examples, usually of papers that catch the attention of the media. Public health researchers and sociologists use the fallacy a lot. And so does the government.

Above I used an example of PM2.5. As far as I have been able to discover, the usual practice is to rely on proxies and *not* actual exposure of PM2.5. A typical abuse is to measure an average value of PM2.5 or to posit a "land-use model" of PM2.5 and then suppose everybody in or near some locale was exposed to the predicted level of particulates. The uncertainty in these models or guesses is *never*—not once that I could discover—carried forward. You might think such egregious practices are rare. You would be mistaken. Indeed, I have yet to see a paper which claims PM2.5 causes disease that didn't rely on the epidemiologist fallacy. For examples see e.g. [60, 113, 125, 171, 172, 181, 203]. See also [86] for a discussion of what this over-confidence means in this application.

I would rather not point out specific studies because this will make it appear the problem is less pervasive than it actually is. This is because if I mention one study, another similar will get a pass because it wasn't mentioned. I also don't want to hurt the feelings of any authors. Mostly, their gross misuse of statistics is not their fault. It was how they were taught to do it. But I am obliged to give some idea of the extent of the problem. So here are some examples.

The study [231] "A Population-Based Case-Control Study of Extreme Summer Temperature and Birth Defects" purportedly investigated birth defects in New York residents (the Y) and heat waves during pregnancy (X), which were claimed to increase in frequency and severity once global warming finally strikes. "We found positive and consistent associations between multiple heat indicators during the relevant developmental window and congenital cataracts [in newborns]". Various statistical measures of correlation were attested to, and if the reader wasn't careful she would decide to stay out of the heat lest her unborn child develop congenital cataracts.

But exposure of women to heat during their "relevant development windows" was never measured on any woman. There was no X. But there was a (let's call it) W said to be the X: the daily air temperature at "18 first-order airport weather stations". Women were assigned the temperature at the stations closest to where they listed their residence at the time of birth for just those days thought to be crucial to fetal development. Nobody knows where the women actually were during these days: it may have been near the assigned airport, or it could have been Saskatchewan, or perhaps in some cool building. The authors also say "we were unable to incorporate air conditioner use data". This paper was taken seriously by the press.

My personal favorite is the paper by David Yanagizawa-Drott and Andreas Madestam [154] who suspected Fourth of July parade attendance (X) turned innocent Americans into Republicans (their Y). How? Exposure to raw, unfiltered patriotism would take its inevitable toll and cause people to turn wistful at the mention of Ronald Reagan. They speculated, "Fourth of July celebrations in the United States shape the nation's political landscape by forming beliefs and increasing participation, primarily in favor of the Republican Party." The more parades attended, the greater the likelihood of turning Republican.

It was very widely reported that X caused Y. Only it wasn't so. The authors never measured X. Instead, they gathered precipitation data from 1920 to 1990 in towns where study participants claimed to have lived when young. If it rained on the relevant Fourths of July, the authors claimed the participants did not go to a parade, because they assumed all parades would be canceled. If it did not rain, they claimed participants did go to a parade, because (they assumed) all towns invariably had parades on clear days, and if there was a parade participants attended. Nowhere was actual parade attendance (X) measured. Were any participants at grandma's house on those weekends instead of home? And think: if their hypothesis is true, San Francisco would be teeming with Republicans because it almost never rains there on the Fourth of July.

The epidemiologist fallacy is only a start. Not only do scientists incorrectly say that they measured an X when they did not, sometimes they do not even measure the Y and accept a proxy for it, too. Yet the story is still X causes Y. Sociologists have the most fun with this. Take the article "Red Light States: Who Buys Online Adult Entertainment?" by Benjamin Edelman [66] who claimed red states consume more pornography than blue states with the strong implication that conservatives are naughtier than progressives. Yet no individual's consumption or political views were ever measured.

I beg the reader to go to the literature and confirm for himself how widespread the epidemiologist fallacy is. Anything focused on human behavior will do, including especially politically sensitive subjects.

10.5 Quantifying the Unquantifiable

The devastation to sound argument by quantifying the unquantifiable cannot be quantified. But it is monumental. Every time a survey with arbitrarily quantified answers is presented as a discovery or as "confirmation" of common sense the road to scientism widens. Here is simple proof of this claim.

On a continuous scale of -3.2 to $113^{1/3}$, how would you rate the keenness of quantifying the unquantifiable? Or how about something more scientific: on a scale of 1 to 8.3, in units of $1/r$ where r is a prime number, how flummoxed does the previous question make you? Do you think that a "flummoxedness" of 8 is twice as flummoxed as a "flummoxedness" of 4? And is a "flummoxedness" of 4 twice as flummoxed as a "flummoxedness" of 2? Have we captured all there is to know about "flummoxedness" in this "scientifically validated instrument"?

Groups of questions are often called "instruments" in order to mimic the prestige things like x-ray spectroscopy have. Social "instruments" are questions with *ad hoc* numerical values attached to them, given first to a group of volunteers (likely college students since the designers are mostly professors), and then given again to a second group. *Ad hoc* scores are created from the quantified answers; sometimes the number of separate scores calculated from even a limited set of questions can be quite large. If the scores are somewhat similar between the first and second group, the scientist calls his "instrument" "validated." It is then released into the wild—though often with a copyright attached so that would-be administrators of the instrument are made to pay for its use.

There is an old Russian proverb which says the mind of another is like a black forest. Yet modern science thinks it can plumbs the depths of any soul if only enough quantified questions are asked. Though researchers say they are careful about question wording, ambiguity in language, particularly about emotion states and highly charged political questions is high. There is always the danger of questions designed to solicit the desired outcome. Consider "Are you in favor of helping the destitute or would you rather they just die and decrease the surplus population?" versus "Are you in favor of a massive tax hike to fund a new inefficient stream of welfare?"

Nobody disputes that there are levels of, say, happiness. One can be amused, pleased, gratified, sated, satisfied, gleeful, ecstatic, serene, gloomy, depressed, sad, grieved, aggrieved, and on and on. Yet it is only hubris that allows a researcher to say, "How happy are you on a scale from 1 to 10?" and think he has well quantified this complex emotion merely because some people checked off a number. But even this might be okay, this crude, blundering quantifying of the unquantifiable—after all, this is the purpose of all those different words we have for happiness—except that the research *must* go on and submit his answers to classical statistical analysis. Calamitous over-certainty is the result.

Quantifying the unquantifiable got its real start with Rensis Likert's (then at New York University) original 1932 paper "A technique for the measurement of attitudes" [148]. Studying this work is revealing. In that paper, Likert (pp. 9–10) said "A series of verbal propositions dealing with the same general social issue are assumed to be more or less equivalent, or at least to be closely related so as to permit prediction from a knowledge of a subject's attitude on one issue to the same subject's attitudes on other aspects of the same issue. . ... In statistical language, a *group factor* is assumed at the outset." He was concerned with measuring "pro- or anti-Negro feeling" on several questions which were to be combined into a "Negro scale" (and a separate "Internationalism scale" which I'll ignore here). Initiating a practice which would become customary (p. 14) "the attitudes tests were given to undergraduates (chiefly male) in nine universities and colleges." There were 15 questions fed into his "Negro scale", each question receiving different numerical weights. Some questions were scaled (in order), such as "Yes", "?" (he meant indeterminate), or "No"; others were multiple choice. Here is question eight, scored 5 points for (a) down to 1 point for (e):

In a community where negroes outnumber the whites, a negro who is insolent to a white man should be:

(a) excused or ignored.
(b) reprimanded.
(c) fined and jailed.
(d) not only fined and jailed, but also given corporal punishment (whipping, etc.).
(e) lynched.

Is it really the case that because lynching an "insolent negro" is worth only 1 point and that fining and jailing him is worth 3? Excusing insolence is only five times as merciful as lynching? The answers are obviously not equally spaced in their emotional or cultural content; not for us and not for his nine (yes, nine) students in 1932.

Here is question 12: "If the same preparation is required, the negro teacher should receive the same salary as the white", with answers (values) "Strongly approve" (5), "Approve" (4), "Undecided" (3), "Disapprove" (2), and "Strongly disapprove" (1). Is it the case that strongly disapproving that a "negro" should receive the same pay as a white teacher is *numerically equivalent* to lynching a man? It *must* be, because Likert's "Negro scale" is a simple average of the answers from all questions, including this one. Except for some adjustment because some questions allowed differing numbers of answers, all questions in the scale are weighted equally. Thus lynching and strong disapproval of equal salaries are by definition *morally equivalent.*

The percent of answers given to question eight, for example, were: (a) 29 %, (b) 42 %, (c) 26 %, (d) 3 %, (e) 0 %. Perplexingly, Likert said this and similar responses "yielded a distribution resembling a normal distribution." That is so only if we take "resembling" in the same sense as "a man resembles a snail" because both are animals. He used this approximation of normality to develop what he called "Sigma scoring" (which is not of direct interest to us) which allowed comparing values of answers of questions with differing number of responses. He summed up the answers, which became the scale.

Likert gave his questions to eight groups of varying sizes (30–100); naturally, the distribution of scores did not match place to place. Variability is expected. But neither did the scores from the same places match the scores on a re-test given 30 days later—there was an 0.85 correlation (calculated in the standard way). Of course, it could be that opinions of the respondents changed during this time. Or their attitude towards members of a different race might have remain fixed but their understanding of the question wording change. Or it could be that they paid a different level of the attention from that time to this. Or *et cetera*. This kind of uncertainty in the answers is never, so far as I have been able to discover, accounted for in any analysis which uses scored questions.

"Validity" to Likert (and his followers) is how well the means and standard deviations of the scores match at different locations (or times). But that is obviously circular reasoning. What is real validity? It should be how well the score matches the underlying truth. That would be how well our scale above measured flummoxedness, or how well Likert's score measured the complexities of attitudes towards "negroes". But if our best poets and writers can barely plumb these depths, what arrogance it is to suppose some simple quantified questions can!

As before, it's worse than this. Researchers rarely report the result of one "scale", but often how different scales or instruments match one another. If there is any kind of correlation, the emotion or psychological state claimed to be (exactly measured) by one scale is said to either cause or be caused by the emotion or psychological state claimed to be (exactly measured) by the second. What usually happens is that the two scales have similarly worded questions. This is never recognized.

For example, the very widely used Center for Epidemiologic Studies Depression (CES-D) scale in its short form asks level of agreement to *inter alia* the statements "I felt that everything I did was an effort", "I was bothered by things that don't usually bother me", and "I was happy". And the just-as-common SF-12 (a distillation of the SF-36) Health Survey has the questions *inter alia* "Have you felt calm and peaceful?", "Did you have a lot of energy?", and "Have you felt downhearted and depressed?".

Both of these "instruments" have scores. The 12-question SF-12 claims it can measure *ten* separate dimensions of health! One of these 10 is "vitality". Researchers will model (usually with regression) the relationship between the scores from the CES-D and SF-12, and when "significance" is discovered, they will say something like "Depression lowers vitality" or "Vitality lessens depression." They will then scour the characteristics of the people measured for clues of how to raise vitality and thus lower depression.

The over-certainty of these works is staggering. Has the CES-D really told us all we know about flummoxedness; or, rather, depression? Has the SF-12 measured with crystal precision vitality? Certainly not. Yet it is *always* assumed the scale encapsulates every important facet of the emotion or psychological state assigned. The objection that the wordings of questions are similar between two or more "instruments" is *never* made. The scores are always assumed linear. That magic trick is what allows regression to enter and for "sub-group analysis" to flourish. There are hundreds of standard questionnaires in regular use, which has led to a this-correlated-with-that literature, the very existence of which is used as evidence that researchers know what they are doing. Researchers take comfort that others are doing as they, which is all the proof required that all is well. Once somebody gets a stupendously over-certain theory into print, it is license for more such claims (but this time studied on this or that interesting group that was heretofore "neglected").

Because of the way they are developed, even *inside* questionnaires, the similar-wording objection can be made. Researchers designing an "instrument" conjure long lists of questions thought to be related to the emotion of interest. These questions are given to a test audience, as it were, and the questions are successively winnowed by ascertaining how close answers to each other question match in the test sample. This closeness is taken as proof that the questions are measuring the stated emotion. But does "Do you like the color blue" really differ from, "About blue, rate how good it makes you feel?"

For a contemporary example, consider this hotly controversial topic. The paper "Psychoticism, Immature Defense Mechanisms and a Fearful Attachment Style are Associated with a Higher Homophobic Attitude" by Ciocca et al. [46] was picked up by the press and announced with the headline "New Study Suggests Connections Between Homophobia And Mental Disorders".[2] This press article opened, "Homosexuality was long derided as a mental disorder...but a new study suggests that it might be more likely that it's actually homophobia that is a sign of mental disorder." The article quoted one of the study authors (E.A. Jannini) as saying, "After discussing for centuries if homosexuality is to be considered a disease, for the first time we demonstrated that the real disease to be cured is homophobia, associated with potentially severe psychopathologies." Potentially *severe* psychopathologies? Sounds like the sort of thing that requires treatment, perhaps against the will of patients.

Now this study asked a few hundred Italian students questions from something called the "Homophobia Scale" as "validated by Wright, Adams, and Bernat" ([226]) and more questions from the Symptom Check List-90-R (SCL-90-R), "one of the most widely used self-report psychometric tests in the area of psychopathologic symptom assessment", which provides nine "indexes," one being "psychoticism" (for the SCL-90-R questions, see [176]). A regression was ran from the quantified answers and a wee p-value "confirmed" (notice the fallacious identification of cause) "homophobia" described a "significant predictive value of psychoticism."

Italy is, of course, largely a Catholic country; indeed, 75 % of the respondents identified as Catholic. Faithful Catholics are obliged to hold, and many do hold, natural law views of homosexuality which consider homosexual acts as unnatural, sinful, and harmful or (as the catechism has it) "objectively disordered". Further, while homosexual acts are condemned in the Bible, Catholics are taught to "love the sinner, hate the sin." These views are central Church teachings, yet there is, as all

---

[2]http://thinkprogress.org/lgbt/2015/09/11/3700857/homophobia-mental-disorder/, Accessed 13 September 2015.

know, much variation in what individual Catholics believe. Whether these attitudes are right or wrong is immaterial to the discussion here. What *is* relevant is that Catholic views toward homosexual acts are in part or in whole matters of religion *and* philosophy, and therefore "homophobia" *must* be considered in those contexts. Do the scales relied upon by the researchers account for religion and philosophy? No: they do not even come close.

For proof, here are some of the questions on the 25-item "Homophobia Scale", scored 1–5, "Strongly agree" to "Strongly disagree": "3. Homosexuality is acceptable to me", "8. Marriage between homosexual individuals is acceptable", "12. Homosexuality is immoral", "16. Organizations which promote gay rights are necessary", and "20. Homosexual behavior should not be against the law." Now each of these contribute to the score in the obvious way towards being "homophobic". But each also has an answer which is the opposite of homophobia (in the colloquial sense of, say, "hating gays") but which is in line with Catholic doctrine. We should therefore expect those who are more religious or philosophical to have greater "homophobia" scores but who are not actually homophobic in the sense used by the study authors, which they define as "irrational fear, hatred, and intolerance of homosexual men and women by heterosexual individuals."

In the SCL-90-R there are ten questions related to "psychoticism." The questions are scored from 0 to 4, expressing agreement "Not at all" to "Extremely". Some of these, if answered *honestly and forthrightly* by participants, a condition nobody can know, clearly indicate mental difficulties, to say the least. Two questions indicating what most would consider mental illness: "7. The idea that someone else can control your thoughts" and "16. Hearing voices that other people do not hear". But there are also questions that would just as obviously be answered by religiously or philosophically minded people that do not indicate illness in the context of this study. These are: "84. Having thoughts about sex that bother you a lot", "85. The idea that you should be punished for your sins". Don't forget that these questions were asked immediately *after* the questions on homosexuality, so that any ideas participants had in this direction were likely amplified.

It is thus no surprise that scores from these two "scales" should exhibit rough correlation—which is exactly what was found: a wee *p*-value in a model with a small effect and low proportion of variability explained (as is typical), and where the authors gave no indication of having considered alternate explanations. Yet the lead researcher was able to claim, in public, that "homophobia" is a "disease" "associated with potentially severe psychopathologies." This is not science. It is advocacy or sloppy thinking. There is no third alternative.

We can now see that in the scale of research which causes over-certainty, the runners-up are those researchers whose work leads to newspaper headlines which begin "Science confirms...". These confirmations will be on such things as "Dressing well improves peoples' opinion of you", "Blisteringly hot days are 'perceived' as more comfortable than clement days", "Daytime has more luminosity than nighttime", and so on endlessly. These studies come out with distressing regularity, all driven by researchers' need to publish something, anything, and all of which enhances scientism, the fallacy that they only way to know anything worth knowing is if somebody in a white lab coat has certified it. Incidentally, take a moment to spot the real study among the headlines just given. Have it? It's the first, from the paper "The Cognitive Consequences of Formal Clothing", [199].

The damage done to clear thinking by pretending batteries of questions adequately quantify emotional states cannot scarcely be underestimated.

10.6   Time Series

The mathematical implementation of time series, like in all other published analyses, is usually flawless, but frequent mistakes are made in interpretation, most of which include: (1) substituting models for actual data, i.e. reification; (2) confusion over statistical significance, (3) model choice and verification, (4) mistaking probability for causality, (5) over-smoothing data, (6) using parametric and not predictive error bounds, and (7) counterfactual questions.

Assume y_t, a time series, is measured without error. The simplest, fairest, and grief-free analysis "method" is to summarize or plot the data without manipulation, *sans accoutrement*. A common question is, "Has this time series increased [or decreased]?"; a near equivalent is, "Is there a trend?" Unfortunately, these questions, plain as they seem, are ambiguous. Consider the data in Fig. 10.5, the result of a simple ARMA(1,1) process simulation (using R's `arima.sim()` function; remind yourself exactly what "simulation" means in Chap. 6). As is clear, sometimes $y_{i+k} > y_i$, $i, k = 1, 2, \ldots, n$, sometimes $y_{i+k} < y_i$, and sometimes $y_{i+k} = y_i$.

In this case, "Has the temperature increased/has it a trend?" is poorly posed. The answer will be "Yes" or "No", depending on the values of i and k and on the

Fig. 10.5 A hypothetical time series measured without error, simulated with an ARMA(1,1) process

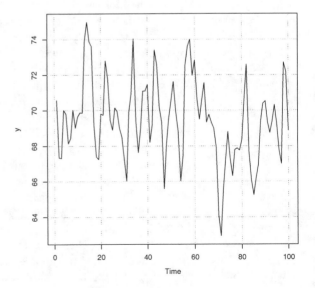

interpretation of the word "increased." *Increase* or *trend* is consonant with "The temperature has gone up more often than it has gone down", "The series is higher at the end than at the beginning", "The arithmetic mean of the latter half is higher than the mean of the first", "The series increased on average at more or less the same rate", "There are more values in the second half (last quarter, or whatever) higher than some constant than in the first half (quarter, etc.)", "Each successive point is equal to or greater than all previous points", "A $w\%$ increase from time i to time $i + k$", and many others.

Whether or not there is an increase, decrease, or trend can be answered without ambiguity or uncertainty, depending only on whether the series has met the criterion specified. One person's increase is another's no change or even decrease. No probability models are needed to distinguish if an increase/decrease/trend is present, and there is no requirement for statistical interpretation. Of course, if the language in the criterion of change is not clear, uncertainty remains. For instance, what exactly does "more or less at the same rate" mean? Obviously, different things to different people.

Authors often write (and the press occasionally repeats) that time-series trends are "statistically significant". There is scarcely any awareness that this "finding" is model and statistic-inside-model dependent (p-values change with different statistics). And even less that it is also conditional on the start and end point of y used. It is enough to claim "significance" that the ritual has been passed. The end point of a time series analysis is almost always picked as y_n, where n is the last observed point. This feels natural; though the temptation to cut the data off at a point more convenient to the author has been noticed. The start point allows for creativity. "Significant" trends appear and disappear by careful choice of beginning. Different impressions of the nature of the series are had by how it is plotted, regardless whether formal testing is invoked.

Figure 10.6 is the same as Fig. 10.5 except that two regression lines, i.e. common models, have been over-plotted. The solid uses all the data, and shows a statistically significant downward trend (effect $= -0.017$ per time unit, t-test p-value of 0.038). The dashed line uses only the data from time 65 onwards and shows a statistically significant upward trend (effect $= 0.076$ per time unit, t-test p-value of 0.027). Depending on the start point, the story one wants to tell changes dramatically.

Most "trends" are identified with straight lines, usually via regression. Yet there isn't any reason to suppose straight-line regressions are good models of reality in most instances. That this above all other models is used reveals the effects of custom.

Now causality. Suppose $y = \{61, 69, 69, 70, 72, 72, 73\}$ (say, yearly average temperature in degrees F; with notation suppressed). What caused the temperature to take the value $y_1 = 61$? The sun, the amount of moisture in the air, especially in the form of clouds, the characteristics of the surface around the thermometer and how it itself is situated, and things of that nature. Not just one cause, but many contributing causes. Most time series are like this in not having a unique cause. And since yearly averages are composed of monthly then of daily averages, which themselves are averages of hourly measurements, it is difficult or impossible to identify clearly all the causes that produced y_1. Regardless of our ignorance, it has some cause or causes. What caused $y_2 = 69$? It wasn't and couldn't have been y_1: y_1 is an *effect*, not a cause. The causes of y_2 were probably similar to the causes of y_1, but they were obviously not the same; if they were the same, then y_2 would equal y_1. And so on for the remaining time points.

Since we can see a "rough" increase in this small set of numbers, defined here as data that contains more increases than decreases, we *know*, with certainty, that some thing or things *caused* this increase. The increase (as defined) is certainly there: we see it. We just don't know why it's there. Suppose our imaginary model suggested the trend was statistically important. This does imply a linear (or other) cause might have been present. But it might imply multiple coincident causes could have been operating.

Assume our model with linear trend predicts $\Pr(y_8 > 73|y_{1...7}, M) = 0.9$ (the 73 temperature figure was chosen because it was decisionable, or otherwise important to somebody). Did the model *cause* the value of y_2 to be higher than y_1, and will it cause y_8 to be larger than y_7? Obviously not. The prediction can, however, be used to verify the goodness of the model. If it turns out $y_8 = 64$, the model did poorly; if $y_8 = 74$, the model did well.

Predicting and explaining can be complementary here as everywhere. Suppose we have two models, one with and one without a trend component, and further suppose $\Pr(y_8 > 73|y_{1...7}, M_{trend}) = 0.9$ and $\Pr(y_8 > 73|y_{1...7}, M_{no\ trend}) = 0.88$. The trend model says higher values are more likely, but not much more likely. Whether 0.9 is much larger than 0.88 is, of course, a matter for the person making decisions based on these predictions. Probabilities important to one person may be irrelevant to another. The reader might think of this as an approximation to a numerical Occam's Razor. The trend model is more complicated, yet predictions without the tend are not very different than with it. This gives evidence that the

Fig. 10.6 Same time series as in Fig. 10.5, but with statistically significant trend lines drawn, each with a different start point. The *solid line* uses all the data and reveals a statistically significant downward trend; the *dashed line* reveals a statistically significant upward trend starting at time point 65. The data were the result of a simple ARMA process simulation

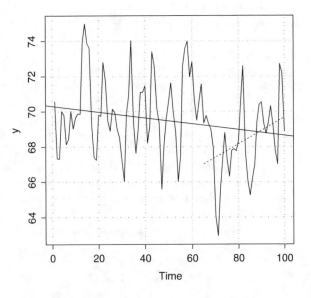

trend is not especially explanatory (but it is not proof). As long as one does not inadvertently insert causal language into the mix, this kind of comparison is immensely useful.

Another popular way to represent time series is like this:

$$y = 67 + \{-6, 2, 2, 3, 5, 5, 6\},$$

where y is supposed to be some central or mean value which is "shocked" into departures. But that implies y returns to, or "desires" to return to this value and is then continually "re-shocked". However useful a modeling concept this is, it is a mistake to think about it in terms of causality. This is to anthromorphize the data or to make a plea to mysterious forces.

When dealing with a single time series, the risk of reification and over-certainty are great when smoothing is applied. (This discussion concerns all types of analyses where the model is shown in preference to reality.) The smoothed data becomes more important than the noisy, uglier reality, and the model which the smoothing implies is more probable than evidence dictates. With two or more series, smoothing artificially increases the correlation between the series, again leading to over-certainty. "Smoothing" is the replacement of the data with a model or with a more aesthetically pleasing substitute, sometimes used to remove "noise". Saying there is "noise" implies one knows the cause. But if one knew the cause, then one needn't smooth.

Figure 10.7 is simulated normal "noise", plotted in light gray, over which appears in black a ten-unit rolling mean, a typical smoother (the example which follows

Fig. 10.7 Generated numbers (in *gray*) with a ten-unit rolling mean (in *black*). The smoothed data gives the false appearance some regular, periodic cause is operating

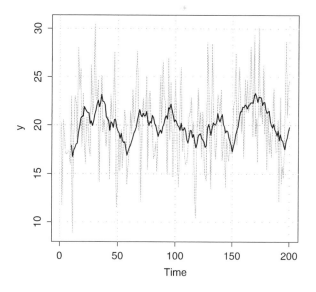

works with any form of smoothing). As explained above, some thing or things *caused* each of the values (in this case a fixed algorithm inside my computer). But the black line gives the appearance that the underlying meta or single cause has been identified and that it has this shape. But if we knew what this cause or these causes were, we would show them and not the data. Smoothing implies a model, and it might be a reasonable model, but since there is no verification of this model's predictions of new data, there is no way to judge how reasonable the smoothing model is. Thus smoothing gives the appearance of greater certainty than there is. Bluntly, smoothing is often a way of faking it, of suggesting more is known than actually is.

Shown in Figs. 10.8 and 10.9 are a succession of images with two simulated normal noise time series per panel that have nothing to do with one another. Figure 10.8 shows the series with the correlation in the titles, and Fig. 10.9 shows the series in x-y plot fashion overplotted with a regression line. The smoothing was produced by increasing the window (k) of running means, but any type of smoothing (e.g. "low-pass filters") will produce similar effects. As the smoothing increases, the correlations increase from near 0 to something quite high (in absolute value). I urge the reader to try it for himself, experimenting with different kinds of smoothers (and not just running means). Some surprising results can be had.

Of course, any given smoothing may decrease (in absolute value) the correlation between two or more series and not increase it. To discover how general any increase smoothing causes would require specifying not only the kind of smoother, but the probabilistic structure of the time series, and so forth; a worthy investigation but one which would take us too far afield here.

Experience shows the danger, however, is real and common. The reason the trick "works" is that smoothing takes uneven points and "straightens" them, making them more line-like, as the plots in Fig. 10.8 show. Any two lines with non-zero slopes have perfect Pearson correlation, as is trivially proved below. *Never* should any time series that has been smoothed be used as input for "downstream" analyses, e.g. that which shows how the time series is associated with some external x. This substitutes fake data for real, and causes massive over-certainty. Yet this mistake is often found. And not only in time series. Regression analyses often make the same error.

Here is another warning that usually goes unheeded: if the data is reconstructed by proxies or is measured with error (or both), the plus-or-minus error bars accompanying the time series should be predictive, not parametric. Using the latter introduces dramatic over-certainty. See also [141] for mistakes made in comparing time series. One form of reconstruction, as it were, is to take data from several locations and use it to interpolate values at unmeasured locations. Another is when a set of locations, possibly changing over time, are used to calculate an index, e.g. "global average temperature."

Figure 10.10 shows a simulated time series y (in solid; an ARMA(0.9,0.8)) and a "proxy" x (dashed) which is simultaneously measured (simulated from a gamma with parameters $(y_i^{1.1},1)$). That is, each y_i is first generated, then the x is simulated using a function of y_i as a parameter to the gamma. By design, these two series

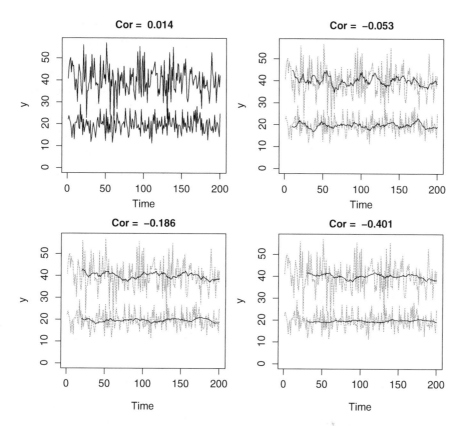

Fig. 10.8 Two simulated normal noise time series, with successively higher amounts of smoothing applied by a rolling mean of k units applied. From top left clockwise: $k = 1, 10, 20, 30$; a $k = 1$ corresponds to no smoothing. The original time series are shown faintly for comparison

closely match. The y series might represent temperature and the proxy might be some oxygen isotope which is used to stand in for y when y is unavailable.

Figure 10.11 is the same as Fig. 10.10, but in a standard x-y plot. From this, it is reasonable to suggest a linear (probability, not causal) relationship between the proxy and time series. I next simulated new proxy data, in the same manner as before: I simulated a new time series and then simulated the proxy from it. This is an oracle approach because we actually know the value of the "unknown" time series. I then assumed that this new proxy data was all I had and wished to use it to estimate the unknown y. I fit a linear regression to the original data in Fig. 10.11, and used it with the new proxy data to predict the new y. In other words, I assumed $y \sim N(\mu, \sigma)$ and modeled $\mu = \beta_0 + \beta_1 \text{proxy}$. Many other models are possible, but this one is not uncommon in the literature, particularly when "trend hunting" or where the "proxy" is replaced by a time model component. What is said below, however, holds for any model.

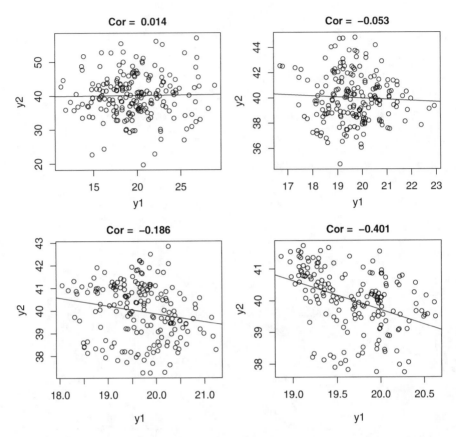

Fig. 10.9 The same as in Fig. 10.8, but the two smoothed time series are plotted against one another and a regression line over-plotted. The effects of the smoothing are quite jarring

The result is in Fig. 10.12: the solid line is the prediction of new y given the new proxy, and the gray envelope is the 95 % parametric confidence interval due to β_1, the parameter associated with the proxy.

The prediction looks good in the sense that, assuming the validity of the model, the confidence limits are fairly tight (interpreted in the natural and not any theoretical way). This is how many proxy reconstructions are plotted. But it is an error to do so, because the confidence in β_1 is of no interest to man or beast. What we want is the *prediction* limits of y, not the uncertainty of the parameter.

Figure 10.13 adds the 95 % *prediction limits* in lighter gray and, because this is an oracle approach and we know the true values of y, these are added in open circles. This is only one simulation among many, but notice that many of the circles are not inside the inner-gray parametric limits, but that most (though not 95 %) are inside the prediction limits.

Note too that the prediction limits are *five to six times wider* than the parametric. Of course, these limits are dependent on the model and the structure of the data, and

Fig. 10.10 A simulated
ARMA(0.9,0.8) series y
(*solid*), and a simulated proxy
x (*dashed*) from a
gamma($y_i^{1.1}$,1)

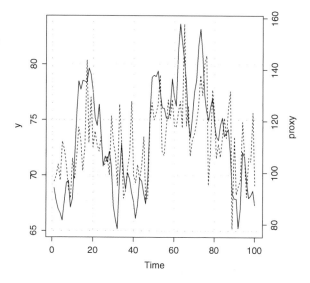

Fig. 10.11 The same as in
Fig. 10.10, but in a standard
x-y plot

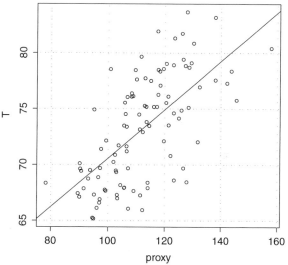

in other applications the difference between the parametric and prediction limits might not be as dramatic. On the other hand, the differences might be, and often are, even larger. One thing is certain, and provable: prediction limits are *always* wider than the parametric intervals, whether one adopts a frequentist or Bayesian stance. Once more, this goes for any kind of analysis, not just time series.

Again, the point is not to investigate strengths or limitations of this regression model, or how the oracle approach reproduces the new simulated time series (experience with the simulations shows that sometimes the prediction is better, sometimes it is worse). Instead we want to understand how to use models in a

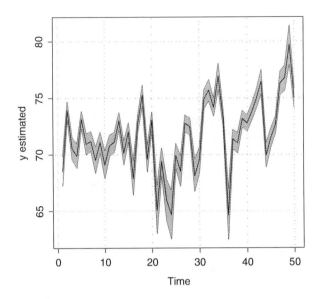

Fig. 10.12 The result of using the regression of proxy on *y* to predict new values of *y* assuming only the proxy is available. The *black line* is the prediction, the *gray envelope* is the 95 % confidence interval of the regression parameter

predictive sense. One point of importance that might be lost is that any model assumes that the proxy-*y* relationship is *unchanged* over the period which the proxy was measured and *y* was not. Since this is subject to some unquantifiable doubt, the quantitative results of uncertainty will themselves be too certain.

The worst mistake researchers make is to use only the solid black line, i.e. the prediction, from Fig. 10.12. Perhaps that prediction is used as input to other "downstream" analyses, or it's just displayed as is. Doing this ignores all uncertainty in the prediction, a bad but frequent habit. This error is almost habitual with some who take (particularly temperature) predictions as input to models of other things (things "caused" by global warming, for instance). The result of ignoring the uncertainty will be to create in the downstream model *p*-values which are much too small or posteriors that are too narrow. Many poor decisions are made.

Slightly better is to incorporate the parametric uncertainty in the same figure (the darker gray envelope), which is rare enough. Best, and most honest, to use the complete predictive uncertainty, and this is mandatory if the predicted series is used as input for outside or future analyses.

Everything said in this section holds for time series measured with error, where the observed data are themselves a proxy for the real, unmeasurable data. In these cases, $w = y + \epsilon$ is measured instead of *y*, where the ϵ represents the departure from the true observation (this needn't be linear error as I have it; any sort can be realized). Many measurement model errors exist (e.g. [37]), the purposes of which

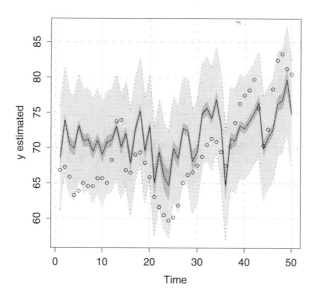

Fig. 10.13 Same as in Fig. 10.13 but with the 95 % prediction limits (in *light gray*) and actual new values of y (*open circles*). The prediction limits are five to six time wider than the parametric limits, a dramatic difference

are to estimate, as above, y. This must also be done in a predictive and not parametric manner.

Particularly in economic and environmental data, external changes are known to have occurred over the span of y. For example, temperatures measured at a fixed geographical location which at one time was an open field but which now lies inside a bustling city. If a town has grown around y, the data are still the data; that is, the y is that still which is experienced.

Some researchers attempt to estimate what y would have been had the external changes not occurred. The estimate is compared against the observed y, and the differences are said to be caused by the external changes. These are interesting, but counterfactual questions. As such, the output produced by any statistical analysis can never be verified. The cause due to the external changes might account for the (model-dependent) differences between the estimate and observation, but then again they might not. There is no way to know. The number of possible causes are vast, and the chance the researcher has hit upon the one-and-only best model in every situation are slight.

These criticisms do not apply to so-called change point analysis, e.g. [44]. This is where external causes to y are sought; or rather, the times at which these changes occurred is sought. This can be a useful procedure if the model is used to make predictions.

Shown in Fig. 10.14 are the monthly temperature anomalies from January 1997 until October 2014, De Bilt, The Netherlands, [118]. This data is interesting because it has been manipulated five times: four times the measurement device changed or

Fig. 10.14 Monthly
temperature anomalies from
January 1997 until October
2014, De Bilt, The
Netherlands

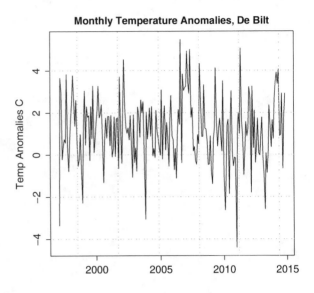

was moved, and a "warming trend of 0.11°C per century caused by urban warming"
was removed. This removal the researchers called "homogenization". As stated
above, this practice is not justified. For one, it's not clear how "stitching" the data
from different measurement sources effects any downstream analysis. Another and
larger problem is that how much urban warming there has been is a counterfactual
question, one to which nobody knows the answer. There must be uncertainty
generated in this homogenization, uncertainty that we should carry forward. I do
not do that here because how to do this would carry us too far afield. The reader
should remember that everything that follows below is thus *too certain*.

Next, "anomalies" were created, a common practice, by subtracting from each
actual temperature the mean of monthly temperatures from the 30-year-period
1961–1990.

What makes the period 1961–1990 special? Nothing at all. Shown in Fig. 10.15
are every possible set of anomalies created with each available 30-year-block of
temperature (the data began in January 1901). The data are now shown as a gray
envelope to indicate the uncertainty inherent in choosing the starting date of the
block.

As is plain, there is tremendous range for the "true" value of any anomaly, up to
2 °C. The decisions one makes based on these values can thus vary remarkably. And
this is only shown for 30-year blocks. What makes 30 years special? Nothing at all.
Increasing the latitude in number of years only increases the envelope of possible
"true" values.

Suppose we define a trend for this series as if the value of the coefficient in a
straight-line regression of the data is non-zero. Does such a trend exist? Figure 10.16
shows the regression lines for every possible set of anomalies created by 30-year

Fig. 10.15 Same as above,
but with every possible set of
anomalies created with
30-year-blocks of temperature

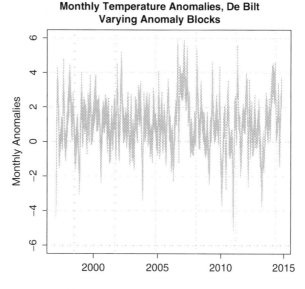

Fig. 10.16 Same as above,
but with regression lines
added for every anomaly
block

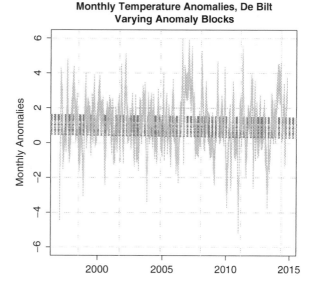

blocks. The modeled decrease per decade was anywhere from 0.11 to 0.08 °C, a fair
discrepancy.

What made starting at 1997 special? Nothing at all. Figure 10.17 is the series
of regression lines one gets starting separately from January 1990 and ending at
December 2012 (so there'd be about two years of data to put into the model) through
October 2014. Solid lines are statistically "significant", dashed "insignificant" (it's
difficult to see at this resolution, but the majority are "significant"). This picture
is brilliant for two reasons, one simple, one shocking. The simple is that we can

Fig. 10.17 Same as above, but with regression lines added for every start point up to December 2012

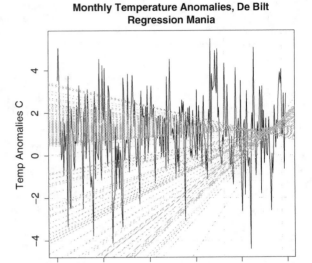

get positive or negative trends by picking various start dates. That means if we're anxious to tell a story, all we need is a little creativity. This picture is just for the 1961–1990 block. Different ones would have resulted if had I used different blocks.

What's shocking about this picture? Researchers will draw a regression line starting from some arbitrary point (like I did) and end at the last point available. This regression line is a model. It says the data should behave like the model; perhaps the model even says the data is caused by the structure of the model (see above).

But the model also logically implies that the data *before* the arbitrary point should have conformed to the model. The start point was arbitrary. The modeler thought a straight line was the thing to do, that a straight line is the best explanation of the data. That means the data that came before the start point should look like the model, too.

Does it? Not hardly. Pick any line, particularly among the increases. Does *all* the data conform to this model? No. Yet each of these models can be (and usually is) considered "correct." The obvious absurdity means the straight line model stinks. Does it matter whether some parameter within that model exhibits a p-value less than the magic number? These models have nothing to do with reality. Even less when we realize that the anomaly block is arbitrary and the anomalies aren't the data and even the data is "homogenized". We could have insisted a different regression line belonged to the period before each of our arbitrary start points, but that sounds like desperation.

Since it often comes up, here is the quick proof two straight lines are perfectly correlated. Pearson correlation r is calculated as:

$$r = \frac{1}{n-1} \sum_{i=1}^{n} \left(\frac{x_i - \bar{x}}{s_x} \right) \left(\frac{y_i - \bar{y}}{s_y} \right) \tag{10.3}$$

and, e.g., $s_x = \sqrt{\frac{1}{n-1} \sum_{i=1}^{n} (x_i - \bar{x})^2}$.

Two straight lines can be written as $x : a + b\{I\}$ and $y : c + d\{I\}$, where $\{I\}$ are the discrete points where the lines are measured. These needn't be these consecutive numbers, but they have to match for x and y, which they will because these are the points at which the two lines are measured (think of any ordinary x-y plot). For ease, suppose $\{I\} = 1, 2, \ldots, n$

First calculate $\bar{X} = \frac{1}{n} \sum (a + bi) = a + b(n + 1)/2$; and similarly for the mean of y. It is also obvious $(n - 1)$ cancels in (10.3). Now $X_i - \bar{X} = b(i - (n+1)/2)$ and $Y_i - \bar{Y} = d(i - (n + 1)/2)$. So $(X_i - \bar{X})^2 = b^2(i - (n + 1)/2)^2$ and similarly for y.

Then the numerator of (10.3) is

$$\sum b(i - (n + 1)/2)d(i - (n + 1)/2) = bd \sum (i - (n + 1)/2)^2$$

and the denominator is

$$\sqrt{b^2(i - (n + 1)/2)^2} \sqrt{d^2(i - (n + 1)/2)^2} = \pm bd \sum (i - (n + 1)/2)^2.$$

We are left with $r = +1$ if b and d have the same sign, or $r = -1$ if b and d have opposite signs. If b or d is 0 the problem is undefined.

10.7   The Future

My prayer is that predictive methods catch on; indeed, that they replace the parameter-centric, hypothesis testing decision-is-probability classical procedures that we now know are so productive of over-certainty. But will these new—actually, conceptually they're rather old—methods be seen as a menace to those who are happy with the present state of affairs, generous as classical methods are in the production of "results"? I am calling for a dramatic slow-down in research: that is the consequence of validating models. Time has to pass before one speaks of having any reasonable certainty in any theory. Who wants a *decrease* in certainty? That is what the logical approach gives. My experience talking to folks about the predictive methods is that it is a hard sell. People *like* the over-certainty provided by classical approaches. Decision making is easy because the software is designed to produce "significance"; and folks don't like the mental effort and emotional turmoil that comes in being less sure. Those two reasons alone account for how classical statistical methods have become so widespread.

The real problem may be in how we teach students probability. Math should, at first, be de-emphasized; philosophy should come first. I've said elsewhere that it's time to stop teaching frequentist methods to all undergraduates, [25]: this admonition ought to be strengthened to all graduates, too. Most experts who teach our subject, and experts are rare in universities where anybody with even modest training is allowed to teach statistics, believe it to be (1) a branch of mathematics, and (2) entirely about empirical propositions. I hope I have shown Probability is properly a branch of philosophy, albeit one that has practical, mathematical aspects. Probability doesn't just apply to the empirical. Too often in classes designed as the only exposure a student will have (this is the bulk of the students), the teacher, thinking he is doing math, insists on that aspect of probability, and gives short shrift to interpretation. The student thus never recalls anything except that his p-value should be wee. Statistics is reduced to ritual, or magic, [94]. This is proved by how people use probability and statistics after they are released into the wild (and by the examples above). But even in those courses designed for probability students, the emphasis is almost entirely on empirical mathematics. Any philosophy taught is incidental. These biases lead to empiricism and scientism.

There is plenty of math to be done, to be sure. Current combinatoric methods have to be scoured and new ones have to be created to produce working methods for discrete, finite predictions. Comparing how these exact algorithms compare to the continuous, infinite approaches will be of lasting benefit, especially in discovering the origin of parameters in common models.

But the real benefit will be a return to a saner and less hyperbolic practice of science, one that is not quite so dictatorial and inflexible, one that is calmer and in less of a hurry, one that is far less sure of itself, one that has a proper appreciation of how much it doesn't know. As of this writing, the situation is dire. For instance, there is much deserved and genuine angst about the "reproducibility crisis" which plagues, it is not generally recognized, those fields which (over-) rely on statistical methods; see e.g. [15, 119, 145, 162, 167] for details on the crisis.

Hypothesis testing, in either its Bayesian or frequentist forms, and regardless of what it is called, is responsible for the bulk of the crisis, as should now be plain. "Statistical significance", as a term or as a goal, should be treated like the ebola virus, i.e. it should be placed in a tightly guarded compound where any danger can be contained and where only individuals highly trained in avoiding intellectual contamination can view it. Second on the Most Wanted list are parameters. It's clear now that the reason parameters became such intense objects of scrutiny, at the expense of reality, is that parameters and testing are nearly conjoined. It's time for parameters to return to their brown paper wrappers and for reality to once again come into view. Mistaking models as reality, or as certain, rounds out the list. There is hardly any recognition how fragile statistical models are.

The predictive approach eliminate all these problems—but not all problems: humans are still human. Yet this solution only became obvious once we understood the true nature of probability, that it is always-conditional, that it is silent on cause, that it is entirely epistemological. Science needs not to return to reality, its true pursuit. And that means chasing after a proper understanding of essence and cause. Nothing else can eliminate over-certainty; nothing else can restore uncertainty to its proper place.

References

1. Adams, E.W.: A Primer of Probability Logic. CSLI Publications, Leland Stanford Junior University, Stanford (1998)
2. Agresti, A.: Categorical Data Analysis. Wiley, New York (2002)
3. Aitchison, J.: Goodness of prediction fit. Biometrika **62**, 547–554 (1975)
4. Ajdukiewicz, K.: Pragmatic Logic. D. Reidel, Boston (1965)
5. Albrecht, A., Phillips, D.: Origin of probabilities and their application to the multiverse. arxiv.org/abs/1212.0953v3 (2014)
6. Ando, T.: Bayesian predictive information criterion for the evaluation of hierarchical Bayesian and empirical Bayesian models. Biometrika **94**, 443–458 (2007)
7. Archer, E., Pavela, G., Lavie, C.J.: The inadmissibility of what we eat in America and NHANES dietary data in nutrition and obesity research and the scientific formulation of national dietary guidelines. Mayo Clin. Proc. **90**(7), 911–926 (2015)
8. Arjas, E., Andreev, A.: Predictive inference, causal reasoning, and model assessment in nonparametric Bayesian analysis. Lifetime Data Anal. **6**, 187–205 (2000)
9. Armstrong, J.S.: Significance testing harms progress in forecasting (with discussion). Int. J. Forecast. **23**, 321–327 (2007)
10. Aumann, R.J.: Agreeing to disagree. Ann. Stat. **4**(6), 1236–1239 (1976)
11. B. Carter, W.H.M.: The anthropic principle and its implications for biological evolution. Philos. Trans. A **310**, 347–363 (1983)
12. Bailey, D.H., Borwein, J.M., Calude, C.S., Dinneen, M.J., Dumitrescu, M., Yee, A.: An empirical approach to the normality of π. Exp. Math. **21**(4), 375–384 (2012)
13. Bartha, P., Johns, R.: Probability and symmetry. Philos. Sci. **68**, S109–S122 (2001)
14. Barzun, J.: From Dawn to Decadence: 500 Years of Western Cultural Life. HarperCollins, New York (2000)
15. Begley, C.G., Ioannidis, J.P.: Reproducibility in science: improving the standard for basic and preclinical research. Circ. Res. **116**, 116–126 (2015)
16. Benenson, F.C.: Probability, Objectivity, and Evidence. Routledge & Kegan Paul, London (1984)
17. Benétreau-Dupin, Y.: Blurring out cosmic puzzles. Philos. Sci. **82**, 879–891 (2015)
18. Berger, J.: Statistical Decision Theory and Bayesian Analysis, 2nd edn. Springer, New York (2010)
19. Berger, J.O., Selke, T.: Testing a point null hypothesis: the irreconcilability of p-values and evidence. JASA **33**, 112–122 (1987)
20. Bernardo, J.M., Rueda, R.: Bayesian hypothesis testing: a reference approach. Int. Stat. Rev. **70**, 351–372 (2002)
21. Bernardo, J.M., Smith, A.F.M.: Bayesian Theory. Wiley, New York (2000)

22. Billingsley, P.: Probability and Measure. Wiley, New York (1995)
23. Blackwell, D., Girshick, M.: Theory of Games and Statistical Decisions. Dover, Mineola (1954)
24. Bonjour, L.: Epistemology. Rowman & Littlefield, Oxford (2002)
25. Briggs, W.M.: It is time to stop teaching frequentism to non-statisticians. arxiv.org/abs/1201.2590 (2013)
26. Briggs, W.M.: On probability leakage. arxiv.org/abs/1201.3611 (2013)
27. Briggs, W.M.: Common statistical fallacies. J. Am. Phys. Surg. 19(2), 678–681 (2014)
28. Briggs, W.M.: The crisis of evidence: why probability & statistics cannot discover cause. arxiv.org/abs/1507.07244 (2015)
29. Briggs, W.M.: The third way of probability & statistics: beyond testing and estimation to importance, relevance, and skill. arxiv.org/abs/1508.02384 (2015)
30. Briggs, W.M., Ruppert, D.: Assessing the skill of yes/no predictions. Biometrics 61(3), 799–807 (2005)
31. Briggs, W.M., Zaretzki, R.A.: The skill plot: a graphical technique for evaluating continuous diagnostic tests. Biometrics 64, 250–263 (2008, with discussion)
32. Brody, T.: The Philosophy Behind Physics. Springer, New York (1993)
33. Burks, A.W.: Chance, Cause, Reason. University of Chicago Press, Chicago (1977)
34. Burnham, K.P., Anderson, D.R.: Model Selection and Multi-Model Inference: A Practical Information-Theoretic Approach, 2nd edn. Springer, New York (2002)
35. Campbell, S., Franklin, J.: Randomness and induction. Synthese 138, 79–99 (2004)
36. Carroll, L.: Symbolic Logic. Macmillan and Company, London (1897)
37. Carroll, R.J., Ruppert, D., Stefanski, L.A., Crainiceanu, C.M.: Measurement Error in Nonlinear Models: A Modern Perspective. Chapman and Hall, London (2006)
38. Carroll, R.T.: The Skeptic's Dictionary: A Collection of Strange Beliefs, Amusing Deceptions, and Dangerous Delusions. Wiley, New York (2003)
39. Cartwright, N.: How the Laws of Physics Lie. Oxford University Press, London (1983)
40. Cartwright, N.: Aristotelian natures and the modern experimental method. In: Inference, Explanation, and Other Frustrations: Essays in the Philosophy of Science. University of California Press, Los Angeles (1992)
41. Castelvecchi, D.: Feuding physicists turn to philosophy for help. Nature 528, 446–447 (2015)
42. Calude, C.S., Chaitin, G.J.: Mathematics: randomness everywhere. Nature 400, 319–320 (1999)
43. Chaitin, G.: Exploring Randomness. Springer, New York (2001)
44. Chen, J., Gupta, A.K.: Parametric Statistical Change Point Analysis: With Applications to Genetics, Medicine, and Finance. Springer, New York (2011)
45. Christensen, R., Utts, J.: Bayesian resolution of the exchange paradox. Am. Stat. 46, 274–276 (1992)
46. Ciocca, G., Tuziak, B., Limoncin, E., Mollaioli, D., Capuano, N., Martini, A., Carosa, E., Fisher, A., Maggi, M., Niolu, C., Siracusano, A., Lenzi, A., Jannini, E.A.: Psychoticism, immature defense mechanisms and a fearful attachment style are associated with a higher homophobic attitude. J. Sex. Med. 12, 1953–1960 (2015)
47. Cohen, J.: Subjective probability. Sci. Am. 197(5), 128–139 (1957)
48. Cohen, J.: The earth is round ($p < .05$). Am. Psychol. 49, 997–1003 (1994)
49. Colquhoun, D.: An investigation of the false discovery rate and the misinterpretation of p-values. R. Soc. Open Sci. 1, 1–16 (2014)
50. Cook, D.B.: Probability and Schrödinger's Mechanics. World Scientific, Singapore (2002)
51. Cox, R.T.: Algebra of Probable Inference. Johns Hopkins University Press, Baltimore (1961)
52. Dawid, A.P.: Present position and potential developments: some personal views. J. R. Stat. Soc. A 147, 278–292 (1984)
53. Dawid, A.P.: Probability, causality, and the empirical world: a Bayes-de Finetti-Popper-Borel synthesis. Stat. Sci. 19, 44–57 (2004)
54. Dawid, A.P., Vovk, V.: Prequential probability: principles and properties. Bernoulli 5, 125–162 (1999)

55. Diaconis, P.: Finite forms of de finetti's theorem on exchangeability. Synthese **36**, 271–281 (1977)
56. Diaconis, P.: A place for philosophy? The rise of modeling in statistical science. Q. J. Appl. Math. **4**, 797–805 (1998)
57. Diaconis, P., Mosteller, F.: The analysis of sequential experiments with feedback to subjects. Ann. Stat. **9**, 3–23 (1981)
58. Diaconis, P., Mosteller, F.: Methods for studying coincidences. J. Am. Stat. Assoc. **84**, 853–861 (1989)
59. Dictionary, O.E.: random, n., adv., and adj. (2015). Oxford University Press, June 2016. Web. 18 Jun 2016. Online; Accessed 4 Dec 2015
60. Dockery, D., Pope, C., et al.: An association between air pollution and mortality in six us cities. N. Engl. J. Med. **329**, 1753–1759 (1993)
61. Dorigo, T.: The Jeffreys-Lindley paradox. Science 2.0. http://www.science20.com/quantum_diaries_survivor/jeffreyslindley_paradox-87184 (2012)
62. Drake, D.: Utility, sensitivity analysis, and cross-validation in Bayesian model-checking (comment to Gelman et al., 1996). Stat. Sin. **6**(4), 760–767 (1996)
63. Draper, D.: Assessment and propagation of model uncertainty. J. R. Stat. Soc. B **57**(1), 45–97 (1995)
64. Draper, D.: New axioms for rigorous Bayesian probability. Bayesian Anal. **3**, 599–606 (2009)
65. Dupré, M.J., Tipler, F.J.: The Cox theorem: unknowns and plausible value. arxiv.org/pdf/math.PR/0611795v1 (2007)
66. Edelman, B.: Red light states: who buys online adult entertainment? J. Econ. Perspect. **23**(1), 209–220 (2009)
67. Einstein, A., Podolsky, P., Rosen, N.: Underdeterminism: Craig and Ramsey. Phys. Rev. **47**, 777–780 (1936)
68. Ellis, G., Silk, J.: Scientific method: defend the integrity of physics. Nature **516**, 321–323 (2014)
69. EPA: Soil screening guidance: technical background document. EPA/540/R95/128. http://rais.ornl.gov/documents/SSG_nonrad_technical.pdf (2015)
70. Feser, E.: Aquinas. Oneworld Publications, Oxford (2009)
71. Feser, E.: The Last Superstition: A Refutation of the New Atheism. St. Augustines Press, South Bend (2010)
72. Feser, E.: Kripke, ross, and the immaterial aspects of thought. Am. Cathol. Philos. Q. **87**, 1–32 (2013)
73. Feser, E.: Existential inertia and the five ways. Am. Cathol. Philos. Q. **85**, 237–267 (2014)
74. Feser, E.: Scholastic Metaphysics: A Contemporary Introduction. Editions Scholasticae, Neunkirchen-Seelscheid (2014)
75. de Finetti, B.: Foresight: its logical laws, its subjective sources. In: Studies in Subjective Probability, 2nd edn. Krieger Publishing Company, Huntington (1980)
76. Fisher, R.: Statistical Methods for Research Workers, 14th edn. Oliver and Boyd, Edinburgh (1970)
77. Fisher, R.: The logic of inductive inference. In: Collected Papers of R.A. Fisher, vol. 2, pp. 271–315. University of Adelaide, Adelaide (1973)
78. Fisher, R.: Statistical Methods and Scientific Inference, 3rd edn. Hafner Press, New York (1973)
79. Franklin, J.: Resurrecting logical probability. Erkenntnis **55**, 277–305 (2001)
80. Franklin, J.: The Science of Conjecture: Evidence and Probability Before Pascal. Johns Hopkins, Baltimore (2001)
81. Franklin, J.: An Aristotelian Realist Philosophy of Mathematics: Mathematics as the Science of Quantity and Structure. Palgrave Macmillan, New York (2014)
82. Franklin, J.: Non-deductive logic in mathematics: the probability of conjectures. In: The Argument of Mathematics, pp. 11–29. Springer, Dordrecht (2015)
83. Freedman, D.: Statistical models and shoe leather. Sociol. Methodol. **21**, 291–313 (1991)
84. Fuchs, C.A.: Interview with a quantum Bayesian. arxiv.org/abs/1207.2141 (2012)

85. Fuchs, C.A., Schack, R.: Qbism and the Greeks: why a quantum state does not represent an element of physical reality. arxiv.org/abs/1412.4211 (2014)
86. Gamble, J.F.: Pm2.5 and mortality in long-term prospective cohort studies: cause-effect or statistical associations? Environ. Health Perspect. **106**(9), 535–549 (1998)
87. Gardner, M.: Weird Water and Fuzzy Logic. Prometheus Books, Amherst (1996)
88. Geisser, S.: On prior distributions for binary trials. Am. Stat. **38**(4), 244–247 (1984)
89. Geisser, S.: Predictive Inference: An Introduction. Chapman & Hall, New York (1993)
90. Gell-Man, M.: The Quark and the Jaguar. St. Martin's Griffin, New York (1995)
91. Gelman, A.: Diagnostic checks for discrete data regression models using posterior predictive simulations. Appl. Stat. **49**(Part 2), 247–268 (2000)
92. Gelman, A.: One more time on Bayes, Popper, and Kuhn. http://www.stat.columbia.edu/~cook/movabletype/ (2005)
93. Gelman, A.: Induction and deduction in Bayesian data analysis. Ration. Mark. Morals **2**, 67–78 (2011)
94. Gigerenzer, G.: Mindless statistics. J. Socio-Econ. **33**, 587–606 (2004)
95. Gneiting, T., Raftery, A.E.: Strictly proper scoring rules, prediction, and estimation. JASA **102**, 359–378 (2007)
96. Gneiting, T., Raftery, A.E., Balabdaoui, F.: Probabilistic forecasts, calibration and sharpness. J. R. Stat. Soc. Ser. B: Stat. Methodol. **69**, 243–268 (2007)
97. Goldberg, S.: When Wish Replaces Thought. Prometheus Books, Amherst (1991)
98. Goldberg, S.: Fads and Fallacies in the Social Sciences. Humanity Books, Amherst (2003)
99. Good, I.: The interface between statistics and philosophy of science. Stat. Sci. **3**, 386–412 (1988)
100. Good, I.J.: Good Thinking: The Foundations of Probability and Its Applications. Univ. of Minn. Press, Minneapolis (1983)
101. Goodman, N.: Fact, Fiction, and Forecast. Bobbs-Merrill, Indianapolis (1954)
102. Groarke, L.: An Aristotelian Account of Induction. Mcgill Queens University Press, Montreal (2009)
103. Groarke, L.: Jumping the gaps: induction as the first exercise of intelligence. In: Philosophical Analysis, Volume 55: Shifting the Paradigm: Alternative Perspectives on Induction, pp. 455–514. De Gruyter, Berlin (2014)
104. Hacking, I.: The Emergence of Probability: A Philosophical Study of Early Ideas about Probability, Induction and Statistical Inference, 2nd edn. Cambridge University Press, Cambridge (2006)
105. Hájek, A.: Mises redux—redux: fifteen arguments against finite frequentism. Erkenntnis **45**, 209–227 (1997)
106. Hájek, A.: Fifteen arguments against hypothetical frequentism. Erkenntnis **70**, 211–235 (2009)
107. Halpern, J.Y.: A counterexample to theorems of Cox and Fine. J. Artif. Intell. Res. **10**, 67–85 (1999)
108. Halpern, J.Y.: Cox's theorem revisited. J. Artif. Intell. Res. **11**, 429–435 (1999)
109. von Hayek, F.A.: The pretence of knowledge. http://www.nobelprize.org/nobel_prizes/economic-sciences/laureates/1974/hayek-lecture.html (1974). Online; Accessed 20 Dec. 2015
110. Hempel, C.: Selected Philosophical Essays. Cambridge University Press, Cambridge (2000)
111. Henricha, J., Heinea, S.J., Norenzayana, A.: The weirdest people in the world? Behav. Brain Sci. **33**, 61–83 (2010)
112. Hersbach, H.: Decomposition of the continuous ranked probability score for ensemble prediction systems. Weather Forecast. **15**, 559–570 (2000)
113. de Hoogh, K., Wang, M., Jerrett, M., et al.: Development of land use regression models for particle composition in twenty study areas in Europe. Environ. Sci. Technol. **47**(11), 5778–5786 (2013)
114. House, J., DeVoe1, S.E., Zhong, C.B.: Too impatient to smell the roses: Exposure to fast food impedes happiness. Soc. Psychol. Pers. Sci. **5**(5), 534–541 (2013)

115. Howson, C., Urbach, P.: Scientific Reasoning: The Bayesian Approach, 2nd edn. Open Court, Chicago (1993)
116. Hume, D.: A Treatise of Human Nature, corrected edn. Oxford University Press, Oxford (2003)
117. Ichikawa, J.J.: The Analysis of Knowledge. The Stanford Encyclopedia of Philosophy. http://plato.stanford.edu/entries/knowledge-analysis/ (2012)
118. Instituut, K.N.M.: De bilt, monthly temperature data. http://knmi.nl/klimatologie/onderzoeksgegevens/homogeen_260/tg_hom_mnd260.txt (2014)
119. Ioannidis, J.P.: Why most published research findings are false. PLoS Med. **2**(8), e124 (2005)
120. Jaki, S.L.: The Limits of a Limitless Science. ISI Books, Wilmington (2000)
121. Jaynes, E.T.: Making Decisions, 2nd edn. Wiley, New York (1985)
122. Jaynes, E.T.: Probability Theory: The Logic of Science. Cambridge University Press, Cambridge (2003)
123. Jeffrey, R.: Subjective Probability. Cambridge University Press, Cambridge (2004)
124. Jeffreys, H.: Theory of Probability. Oxford University Press, Oxford (1998)
125. Jerrett, M.: Spatiotemporal analysis of air pollution and mortality in California based on the American Cancer Society cohort. Contract 06-332. http://www.arb.ca.gov/research/single-project.php?row_id=64805 (2011)
126. Johns, R.: A Theory of Physical Probability. University of Toronto Press, Toronto (2002)
127. Johnson, W.O.: Predictive influence in the log normal survival model. In: Modelling and Prediction: Honoring Seymour Geisser, pp. 104–121. Springer, New York (1996)
128. Johnson, W.O., Geisser, S.: A predictive view of the detection and characterization of influence observations in regression analysis. JASA **78**, 427–440 (1982)
129. Jussim, L., Cain, T.R., Crawford, J.T., Harber, K., Cohen, F.: The Unbearable Accuracy of Stereotypes. Handbook of Prejudice, Stereotyping, and Discrimination, pp. 199–227. Psychology Press, New York (2009)
130. Jussim, L., Harber, K.D., Crawford, J.T., Cain, T.R., Cohen, F.: Social reality makes the social mind: self-fulfilling prophecy, stereotypes, bias, and accuracy. Interact. Stud. **6**, 85–102 (2005)
131. Kerns, G.J., Székely, G.J.: Finite exchangeability. J. Theor. Probab. **19(3)**, 589–608 (2006)
132. Keynes, J.M.: A Treatise on Probability. Dover Phoenix Editions, Mineola (2004)
133. Knuth, D.: Things a Computer Scientist Rarely Talks About. Center for the Study of Language and Information, Stanford (2003)
134. Kreeft, P.: Socratic Logic. St. Augustine's Press, South Bend (2010)
135. Kripke, S.: Wittgenstein on Rules and Private Language. Cambridge University Press, Cambridge (1982)
136. Kullback, S.: Information Theory and Statistics. Dover, Mineola (1968)
137. Kyburg, H.E.: The Logical Foundations of Statistical Inference. D. Reidel Publishing Co., Boston (1974)
138. Kyburg, H.E.: Epistemology and Inference. University of Minnesota Press, Minneapolis (1983)
139. Kyburg, H.E., Smokler, H.E.: Studies in Subjective Probability. Krieger Publishing Co., New York (1964)
140. Lad, F.: Operational Subjective Statistical Methods. Wiley, New York (2001)
141. Lanzante, J.R.: A cautionary note on the use of error bars. J. Climate **18**, 3699–3703 (2005)
142. Leblanc, H. (ed.): Techniques of Deductive Inference. Prentice Hall, Englewood Cliffs (1966)
143. Leblanc, H. (ed.): Statistical and Inductive Probabilities. Dover, Mineola (2006)
144. Lee, J.C., Johnson, W.O., Zellner, A. (eds.): Modelling and Prediction: Honoring Seymour Geisser. Springer, New York (1996)
145. Leek, J.T., Peng, R.: Reproducible research can still be wrong: adopting a prevention approach. Proc. Natl. Acad. Sci. **112**, 1645–1646 (2015)
146. Leslie, J.: The End of the World: The Science and Ethics of Human Extinction. Routledge, London (1998)

147. Levi, I.: Imprecision and indeterminancy in probability judgement. Philos. Sci. **52**, 390–409 (1985)
148. Likert, R.: A technique for the measurement of attitudes. Arch. Psychol. **140**(2), 129–138 (1932)
149. Linde, A.: A brief history of the multiverse. arxiv.org/abs/1512.01203 (2014)
150. List, C.: Craig's theorem and the empirical underdetermination thesis reassessed. Disputatio **7**, 28–39 (1999)
151. Little, R.: Calibrated Bayes: a Bayes/frequentist roadmap. Am. Stat. **60**(3), 1–11 (2006)
152. MacCoun, R., Perlmutter, S.: Blind Analysis as a Correction for Confirmatory Bias in Physics and in Psychology. In: Psychological Science Under Scrutiny: Recent Challenges and Proposed Solutions, pp. 589–612. Wiley, New York (2016)
153. Maddy, P.: Mathematical alchemy. Br. J. Philos. Sci. **37**, 279–314 (1986)
154. Madestam, A., Yanagizawa-Drott, D.: Shaping the nation: the effect of fourth of July on political preferences and behavior in the united states. Working Paper. Http://www.hks.harvard.edu/fs/dyanagi/Research/FourthOfJuly.pdf (2011)
155. Marsaglia, G.: On the randomness of pi and other decimal expansions. Preprint. http://www.yaroslavvb.com/papers/marsaglia-on.pdf (2010)
156. Mayo, D.: Bayesian confirmation philosophy and the tacking paradox in I. Website. http://errorstatistics.com/2013/10/19/bayesian-confirmation-philosophy-and-the-tacking-paradox-in-i/ (2014)
157. Meng, X.L.: Posterior predictive p-values. Ann. Stat. **22**(3), 1142–1160 (1994)
158. Mooij, J.M., Peters, J., Janzing, D., Zscheischler, J., Scholkopf, B.: Distinguishing cause from effect using observational data: methods and benchmarks. arxiv.org/abs/1412.3773 (2014)
159. Murphy, A.H.: Forecast verification: its complexity and dimensionality. Mon. Weather Rev. **119**, 1590–1601 (1991)
160. Murphy, A.H., Winkler, R.L.: A general framework for forecast verification. Mon. Weather Rev. **115**, 1330–1338 (1987)
161. Newman, J.H.C.: The Grammar of Ascent. Image Books, New York (1955)
162. Nosek, B.A., Alter, G., Banks, G.C., et al.: Estimating the reproducibility of psychological science. Science **349**, 1422–1425. (2015)
163. Nuzzo, R.: How scientists fool themselves – and how they can stop. Nature **526**, 182–185 (2015)
164. Oderberg, D.: The Old New Logic. MIT, Cambridge (2005)
165. Oderberg, D.: Real Essentialism. Routledge, London (2008)
166. Pearl, J.: Causality: Models, Reasoning, and Inference. Cambridge University Press, Cambridge (2000)
167. Peng, R.: The reproducibility crisis in science: a statistical counterattack. Significance **12**, 30–32 (2015)
168. Phillies, G.D.: Elementary Lectures in Statistical Mechanics. Springer, New York (2000)
169. Polchinski, J.: String theory to the rescue. arxiv.org/abs/1512.02477 (2015)
170. Polya, G.: Mathematics and Plausible Reasoning, vol. II: Patterns of Plausible Inference, 2nd edn. Oxford University Press, London (1968)
171. Pope, C., Thun, M., et al.: Particulate air pollution as a predictor of mortality in a prospective study of us adults. Am. J. Respir. Crit. Care Med. **151**, 669–674 (1995)
172. Pope, C., Turner, M., Jerrett, M., et al.: Relationships between fine particulate air pollution, cardiometabolic disorders, and cardiovascular mortality. Circ. Res. **116**(1), 108–115 (2015)
173. Popper, K.: The Logic of Scientific Discovery. Hutchinson, London (1959)
174. Popper, K.: Conjectures and Refutations in the Growth of Scientific Discoveries. Routledge, London (1963)
175. Press, S.J.: Subjective and Objective Bayesian Statistics, 2nd edn. Wiley, New York (2003)
176. Prinz, U., Nutzinger, D.O., Schulz, H., Petermann, F., Braukhaus, C., Andreas, S.: Comparative psychometric analyses of the scl-90-r and its short versions in patients with affective disorders. BMC Psychiatry **13**(104), 1–9 (2013)
177. Pritchard, D.: Epistemic luck. J. Philos. Res. **29**, 193–222 (2004)

178. Pritchard, D.: Viritue epsitemology and epistemic luck, revisited. Metaphilosophy **39**, 66–88 (2008)
179. Quine, W.V.: Two dogmas of empiricism. Philos. Rev. **60**, 20–43 (1951)
180. Quine, W.V.: Two Dogmas of Empiricism. Harper and Row, Harper Torchbooks, Evanston (1953)
181. Raaschou-Nielsen, O., Andersen, Z.J., et al.: Air pollution and lung cancer incidence in 17 European cohorts: prospective analyses from the European study of cohorts for air pollution effects (escape). Lancet **14**(9), 813–822 (2013)
182. Ringbauer, M., Duffus, B., Branciard, C., Cavalcanti, E.G., White, A.G., Fedrizzi1, A.: Measurements on the reality of the wavefunction. Nat. Phys. **11**, 249–254 (2015)
183. Robert, C.: On the jeffreys-lindley's paradox. arxiv.org/abs/stat.ME/1303.5973v3 (2013)
184. Robert, C.P.: The Bayesian Choice, 2nd edn. Springer, New York (2001)
185. Ross, S.: A First Course in Probability, 3rd edn. Macmillan, New York (1988)
186. Rucker, R.: Infinity and the Mind: The Science and Philosophy of the Infinite. Bantam, New York (1983)
187. Ruhla, C.: The Physics of Chance. Oxford University Press, Oxford (1992)
188. Sankey, H.: Scientific Realism and the Rationality of Science. Ashgate, Burlington (2008)
189. Satel, S., Lilienfeld, S.O.: Brainwashed: The Seductive Appeal of Mindless Neuroscience. Basic Books, New York (2015)
190. Savage, L.: The Foundations of Statistics. Dover, Mineola (1972)
191. Schervish, M.: A general method for comparing probability assessors. Ann. Stat. **17**, 1856–1879 (1989)
192. Schramm, A.: Evidence, hypothesis, and grue. Erkenntnis **79**(3), 571–591 (2014). doi:10.1007/s10670-013-9524-6. http://dx.doi.org/10.1007/s10670-013-9524-6
193. Scruton, R.: Modern Philosophy. Penguin, New York (1994)
194. Senn, S.: Dicing with Death: Chance, Risk and Health. Cambridge University Press, Cambridge (2003)
195. Senn, S.: You may believe you are a Bayesian but you are probably wrong. Ration. Mark. Morals **2**, 48–66 (2011)
196. Senn, S.: Seven myths of randomisation in clinical trials. Stat. Med. **32**, 1439–1450 (2013)
197. Shafer, G.: A Mathematical Theory of Evidence. Princeton University Press, Princeton (1976)
198. Shimony, A.: Bell's Theorem. The Stanford Encyclopedia of Philosophy. http://plato.stanford.edu/archives/win2013/entries/bell-theorem/ (2013)
199. Slepian, M.L., Ferber, S.N., Gold, J.M., Rutchick, A.M.: The cognitive consequences of formal clothing. Soc. Psychol. Pers. Sci. **6**, 1–8 (2015)
200. Smith, R.L.: Bayesian and Frequentist Approaches to Parametric Predictive Inference (with Discussion). Bayesian Statistics, vol. 6, pp. 589–612. Oxford University Press, Oxford (1998)
201. Solomonoff, R.: A formal theory of inductive inference. Part I. Inf. Control **7**, 1–22 (1964)
202. Spiegelhalter, D.J., Best, N.G., Carlin, B.: Bayesian measures of model complexity and fit. Noûs **64**, 583–639 (2002)
203. Stieb, D., Chen, L., Jerrett, M., et al.: Associations of pregnancy outcomes and pm2.5 in a national Canadian study. Environ. Health Perspect. (2015)
204. Stove, D.: Probability and Hume's Inductive Scepticism. Clarendon, Oxford (1973)
205. Stove, D.: Popper and After: Four Modern Irrationalists. Pergamon Press, Oxford (1982)
206. Stove, D.: The Rationality of Induction. Clarendon, Oxford (1986)
207. Stove, D.: The Plato Cult and Other Philosophical Follies. Basil Blackwell, Cambridge (1991)
208. Stove, D.: On Enlightenment. Transaction Publishers, New York (2002)
209. Strevens, M.: Inferring probabilities from symmetries. Noûs **3**(22), 231–246 (1998)
210. Sweeting, T., Kharroubi, S.: Application of a predictive distribution formula to Bayesian computation for incomplete data models. Stat. Comput. **15**, 167–178 (2005)
211. Tetlock, P.E.: Expert Political Judgement. Princeton University Press, Princeton (2005)
212. Thatcher, A.: Relationships between Bayesian and confidence limits for predictions. J. R. Stat. Soc. Ser. B **26**, 176–210 (1964)

213. Tuyl, F., Gerlach, R., Mengersen, K.: Posterior predictive arguments in favor of the Bayes-Laplace prior as the consensus prior for binomial and multinomial parameters. Bayesian Anal. **4**(1), 151–158 (2009)
214. Vilenkin, A.: Many Worlds In One. Hill and Wang, New York (2006)
215. von Plato, J.: Creating Modern Probability. Cambridge University Press, Cambridge (1998)
216. Wallace, W.A.: The Modeling of Nature. CUA Press, Washington, DC (1996)
217. Walley, P.: Statistical Reasoning with Imprecise Probabilities. Chapman & Hall, London (1990)
218. Watson, L.: Nigel short: 'Girls just don't have the brains to play chess'. The Telegraph. http://www.telegraph.co.uk/culture/chess/11548840/Nigel-Short-Girls-just-dont-have-the-brains-to-play-chess.html (2015)
219. Weisstein, E.W.: Random distribution. http://mathworld.wolfram.com/RandomDistribution.html (2015). Accessed 11 Nov. 2014
220. West, M.: Bayesian model monitoring. J. R. Stat. Soc. B **48**(1), 70–78 (1986)
221. Wikipedia: List of chess grandmasters. https://en.wikipedia.org/wiki/List_of_chess_grandmasters (2015). Online; Accessed 22 April 2015
222. Wilhelmsen, F.D.: Man's Knowledge of Reality: An Introduction to Thomistic Epistemology. Prentice-Hall, Englewood Cliffs (1956)
223. Wilks, D.S.: Statistical Methods in the Atmospheric Sciences, 2nd edn. Academic, New York (2006)
224. Williams, D.: The Ground of Induction. Russell & Russell, New York (1947)
225. Williamson, J.: In Defense of Objective Bayesianism. Oxford University Press, Oxford (2010)
226. Wright, L.W., Adams, H.E., Bernat, J.: Development and validation of the homophobia scale. J. Psychopathol. Behav. Assess. **21**, 337–347 (1999)
227. Youdon, W.: Enduring values. Technometrics **14**, 1–11 (1972)
228. Zabell, S.L.: The rule of succession. Erkenntnis **31**, 283–321 (1989)
229. Ziliak, S.T.: Balanced versus randomized field experiments in economics: why W. S. Gosset aka Student matters. Rev. Behav. Econ. **1**, 167–208 (2014)
230. Ziliak, S.T., McCloskey, D.N.: The Cult of Statistical Significance. University of Michigan Press, Ann Arbor (2008)
231. Zutphen, A.R.V., Lin, S., Fletcher, B.A., Hwang, S.A.: A population-based case-control study of extreme summer temperature and birth defects. Environ. Health Perspect. **120**, 1443–1449. (2012)

Index

Printed in the United States
By Bookmasters